Recent Advances in Homeostasis

Edited by Gaffar Sarwar Zaman

Published in London, United Kingdom

Recent Advances in Homeostasis
http://dx.doi.org/10.5772/intechopen.100954
Edited by Gaffar Sarwar Zaman

Contributors
Ladan Mafakher, Elham Rismani, Nasrin Haghighi-Najafabadi, Behzad Shahbazi, Babak Elyasi Far, Rajiv Kumar, Neelam Chhillar, Dhruv Sanjay Gupta, Ginpreet Kaur, Shailey Singhal, Tanya Chauhan, Sandeep Mittan, Moacir Fernandes de Godoy, Michele Lima Gregório, Cato Drion, Gaffar Sarwar Zaman

First published in London, United Kingdom, 2023 by IntechOpen
IntechOpen is the global imprint of INTECHOPEN LIMITED, registered in England and Wales, registration number: 11086078, 5 Princes Gate Court, London, SW7 2QJ, United Kingdom
Printed in Croatia

British Library Cataloguing-in-Publication Data
A catalogue record for this book is available from the British Library

Additional hard and PDF copies can be obtained from orders@intechopen.com

Recent Advances in Homeostasis
Edited by Gaffar Sarwar Zaman
p. cm.

This title is part of the Physiology Book Series, Volume 21
Topic: Human Physiology
Series Editor: Tomasz Brzozowski
Topic Editor: Kunihiro Sakuma
Associate Topic Editor: Kotomi Sakai

Print ISBN 978-1-80355-477-8
Online ISBN 978-1-80355-478-5
eBook (PDF) ISBN 978-1-80355-479-2
ISSN 2631-8261

For EU product safety concerns:
IN TECH d.o.o., Prolaz Marije Krucifikse Kozulić 3, 51000 Rijeka, Croatia,
info@intechopen.com or visit our website at intechopen.com.

IntechOpen Book Series
Physiology
Volume 21

Aims and Scope of the Series

Modern physiology requires a comprehensive understanding of the integration of tissues and organs throughout the mammalian body, including the cooperation between structure and function at the cellular and molecular levels governed by gene and protein expression. While a daunting task, learning is facilitated by identifying common and effective signaling pathways mediated by a variety of factors employed by nature to preserve and sustain homeostatic life. As a leading example, the cellular interaction between intracellular concentration of Ca+2 increases, and changes in plasma membrane potential is integral for coordinating blood flow, governing the exocytosis of neurotransmitters, and modulating gene expression and cell effector secretory functions. Furthermore, in this manner, understanding the systemic interaction between the cardiovascular and nervous systems has become more important than ever as human populations' life prolongation, aging and mechanisms of cellular oxidative signaling are utilised for sustaining life. Altogether, physiological research enables our identification of distinct and precise points of transition from health to the development of multimorbidity throughout the inevitable aging disorders (e.g., diabetes, hypertension, chronic kidney disease, heart failure, peptic ulcer, inflammatory bowel disease, age-related macular degeneration, cancer). With consideration of all organ systems (e.g., brain, heart, lung, gut, skeletal and smooth muscle, liver, pancreas, kidney, eye) and the interactions thereof, this Physiology Series will address the goals of resolving (1) Aging physiology and chronic disease progression (2) Examination of key cellular pathways as they relate to calcium, oxidative stress, and electrical signaling, and (3) how changes in plasma membrane produced by lipid peroxidation products can affect aging physiology, covering new research in the area of cell, human, plant and animal physiology.

Meet the Series Editor

Prof. Dr. Thomas Brzozowski works as a professor of Human Physiology and is currently a Chairman at the Department of Physiology and is V-Dean of the Medical Faculty at Jagiellonian University Medical College, Cracow, Poland. His primary area of interest is physiology and pathophysiology of the gastrointestinal (GI) tract, with a major focus on the mechanism of GI mucosal defense, protection, and ulcer healing. He was a postdoctoral NIH fellow at the University of California and the Gastroenterology VA Medical Center, Irvine, Long Beach, CA, USA, and at the Gastroenterology Clinics Erlangen-Nuremberg and Munster in Germany. He has published 290 original articles in some of the most prestigious scientific journals and seven book chapters on the pathophysiology of the GI tract, gastroprotection, ulcer healing, drug therapy of peptic ulcers, hormonal regulation of the gut, and inflammatory bowel disease.

Meet the Volume Editor

Dr. Gaffar Sarwar Zaman has an MD in Biochemistry from Assam Medical College and Hospital, Dibrugarh University, India. He completed a Fellowship in Diabetes (FID) at Royal Liverpool Academy, UK, a Fellowship in Applied Nutrition (FIAN) at Medvarsity, Apollo Hospitals, India, and a Post Graduate Diploma in Clinical Research (PGDCR) at Symbiosis University, India. He has almost 18 years of experience at King Khalid Government University, Saudi Arabia, and Rajiv Gandhi University of Health Sciences, India. He has expertise in quality development, curriculum design, and e-learning methods. He is also a skilled researcher and a member of the Research and Scientific Committee of the College. He has almost seventy research publications in both national and international journals to his credit. He has also published several books, including Quality Control in Laboratory, Ultimate Guide to Insulin, and Ultimate Guide to Outpatient Care.

Contents

Preface

Medical science has undergone a vast change in the last 25 years. Although a good design with properly prepared drawings and specifications is essential for homeostasis, various reports on medical/laboratory fallacies/misconceptions have raised the visibility of the need to augment homeostasis, which is the basis for most projects and medical educational institutes. Prominence has moved from simply diagnosing and treating disease to identifying and controlling disease, risk factors, and maintaining health, all of which are related to homeostasis initially. A physiologist named Walter Cannon coined the term homeostasis in 1926 in order to clarify what fellow physiologist Claude Bernard had termed the 'inner environment' in 1865. As a result of changes in the external environment, the body's many functions operate in a narrow range of equilibrium, called dynamic equilibrium. Extracellular fluid surrounding the individual cells of the body changes with changes in the external environment, but a narrow range must be maintained to prevent cell, tissue, and organ death. As a result of biochemical reactions that occur within cells, homeostasis is observable. An enzyme's ability to function optimally in the environment of the cell depends on the regulation of pH, temperature, oxygen, ions, and blood glucose concentrations, as well as control of waste products in order to prevent disruptions of the internal environment. Depending on the internal environment, a cell will remain alive and continue to function as long as it serves a functional purpose in its tissue. Setting points, receiving feedback, and regulating homeostasis are key components of homeostasis. To detect and respond to disturbances caused by disruptors, the body has thousands of control systems. In developing a homeostatic control system, this value is used to determine the output. The regulation of homeostasis involves both local (paracrine and autocrine) and reflex (neuroendocrine and nervous system) controls. The body's homeostasis is dependent upon every organ system. A body's integumentary, musculoskeletal, and cardiovascular systems work together to control body temperature; no single organ system operates alone. This book provides a comprehensive understanding of homeostasis and its importance to the health of the human body.

I sincerely hope that this book will enhance medical care and improve patient quality of life. This work would not have been possible without the support of IntechOpen. I am especially indebted to Publishing Process Manager Elena Vracaric and Commissioning Editor Mirena Calmic who have been supportive of my book and who worked actively to provide me with all facilities. I am grateful to all of those with whom I have had the pleasure to work on this project. The chapter authors have provided me with extensive professional help and obliged me on this scientific venture wherever possible. I would especially like to thank Dr. Fernandez de Godoy Moacir, Dr. Drion Cato, Dr. Kumar Rajiv, and Dr. Mafakher Ladan. Nobody has been more important to me in the pursuit of this project than my family. I would like to thank my late parents, Taufiquz Zaman and Jaiba Zaman, whose love and guidance are with me in whatever I pursue. Most importantly, I wish to thank my loving and supportive wife, Mrs. Jarin Tanwir Hussain, who is my ultimate role model, and my two wonderful sons, Naushad Muntasir Zaman and Umar Sarwar Zaman, who provided me with unending inspiration during the project.

Dr. Gaffar Sarwar Zaman
Department of Clinical Laboratory Sciences,
College of Applied Medical Sciences,
King Khalid University,
Abha, Kingdom of Saudi Arabia

Section 1

Introduction

Chapter 1

Introductory Chapter: Homeostasis – A Brief Description and Scope for Recent Advances in the Medical Field

Gaffar Sarwar Zaman

1. Introduction

Homeostasis is described as the balanced stage in a living system (internal chemical physical and social state or conditions) that does not change with time or negligibly [1]. As a result of a certain set of conditions, the organism has optimal functioning, including keeping certain variables within a certain range (homeostatic range). Moreover, there are other variables that need to be regulated despite changes in environment, diet, or activity level, such as the pH of extracellular fluid, sodium, potassium, and calcium ions concentrations, and blood sugar levels. A single regulator or homeostatic mechanism controls each of these variables, which together maintain life.

A state of homeostasis occurs when optimum conditions prevail [2]. The central motivation for organic action is equilibrium, which is maintained by many regulatory mechanisms. There are three parts to all homeostasis control mechanisms: receptors, control centers, and effectors [3]. In addition to monitoring and responding to changes in the environment, the receptor is also responsible for sensing changes within the body. A thermoreceptor and a mechanoreceptor are examples of receptors. Respiratory centers and renin-angiotensin systems are two of the control centers. In order to change back to a normal state, an effector must be acted upon. A nuclear receptor causes gene expression to change at the cellular level by upregulating or downregulating and acts as a negative feedback device. Controlling bile acids in the liver is one example [4].

Homeostasis was coined by Walter Bradford Cannon in 1926 after English physiologist Claude Bernard defined it in 1849 [5, 6]. In 1932, British physiologist Joseph Barcroft proposed that the most stable internal environment was necessary for higher brain function. Barcroft interpreted homeostasis as a service to the brain, rather than one that it organized [7]. The concept of homeostasis refers to the constant environment in which cells live and survive within the body, as described by Bernard and Cannon [8]. Homeostasis is defined as the state of being in balance on a physiological level, but cybernetics describes technological control systems such as thermostats that act as homeostasis but are often defined much more broadly than biological homeostasis [9–12].

All organisms are dependent on very specific chemical and physical environments for their metabolic processes to take place. Chemical processes vary by organism

IntechOpen

and depending on whether they occur inside or outside the cells. Humans and other mammals possess one of the most widely known homeostatic mechanisms which are responsible for maintaining the constant composition of the extracellular fluid (also known as the internal environment). These mechanisms control the temperature, pH, osmolality, sodium, potassium, glucose, carbon dioxide, and oxygen concentrations in the extracellular fluid (or internal environment). However, a great many other homeostatic mechanisms, encompassing many aspects of human physiology, control other entities in the body. Where the levels of variables are higher or lower than those needed, they are often prefixed with hyper- and hypo-, respectively such as hyperthermia and hypothermia or hypertension and hypotension.

In the morning, the body temperature rises to about 37.5°C, then falls to about 36.4°C in the evening, before again rising to approximately 37.5°C in the morning. Having homeostatic control does not mean the value of an entity is always constant in health. The hypothalamus, among others, regulates the core body temperature via a homeostatic mechanism [13]. Regular resets are performed on the regulator's set point, however [14]. People's core body temperatures vary throughout the day (a circadian rhythm), with a low temperature at night and a high temperature in the afternoon. A woman's menstrual cycle can also cause variations in temperature [15, 16]. Infections cause a fever by resetting the temperature regulator [17]. Through the process of acclimatization, organisms can adjust somewhat to different conditions such as temperature or oxygen level changes. Body activities are not all governed by homeostasis [18, 19].

In order to convey information about the direction and magnitude of the error detected by the sensor, the signal from the sensor to the effector is, of necessity, highly variable [11, 20, 21]. A similar mechanism is needed to reverse the effector's error—in fact, it should be in proportion (but in the opposite direction) to the error that threatens the environment within [12]. During the growth of the internal carotid arteries, stretch receptors in the aortic arch and carotid sinuses measure arterial blood pressure [13]. Blood pressure sensor signals are sent through sensory nerves to the medulla oblongata of the brain according to whether it has dropped or risen. Motor or efferent nerves of the autonomic nervous system then carry messages from the medulla oblongata to a variety of effector organs, whose activity is changed accordingly to correct the blood pressure error. When arterial blood pressure drops, the heart's rate increases (tachycardia), and when pressure rises above the set point, it slows down (bradycardia) [13]. As a result, the heart rate (for which the body does not have a sensor) does not function as a homeostatic control, but rather as an effector response to arterial blood pressure errors. Sweating is another example of sweating regulated by the hypothalamus of the brain. The hypothalamus is equipped with a sensor for measuring this effector, which helps to control body temperature by regulating heat load at home. Recent advances in understanding body weight homeostasis in humans.

There is no consensus framework for body weight homeostasis today, despite the assumption of control over body weight. The set point of body weight suggests that (i) it is more or less tightly controlled and (ii) it is symmetric or asymmetrically controlled biologically. This is the result of feedback loops between peripheral organs and tissues (e.g. leptin secreted by adipose tissue) and a central system. It is also possible that metabolic adaptations to energy imbalance occur without feedback control by reaching a "settling point" rather than a set point. An alternative method that combines both paradigms is the "dual intervention point" model, which introduces two set points and a settling point between them. Biological control of body weight is not consistently demonstrated in observational studies on large populations, which may be overridden by obesogenic environments and cultures that affect behavior and experiences. A focused protocol based on sound principles is needed to address the issue of weight homeostasis, such as examining lean rather than overweight subjects before, during, and after weight loss. The association between the mass of individual body components (i) and mass and metabolic function (ii) in contexts of cerebrohumoral control and systemic effects need to be addressed with improved methods and a multi-level–multi-systemic approach. A greater effort should be made to avoid simplifications and non-biologic phenotypes (e.g. body mass index and waist circumference). Control (or a set point) will have more to do with energy expenditure than body weight, as changes in body weight are the result of mismatches between tightly controlled energy expenditure and loosely controlled energy intake [22]. Overall, body weight control is currently offered in three different forms. It is true that all models have limitations and cannot fully explain human weight fluctuations, despite their striking appearance. A tight control system might be invalidated in the short term due to the lack of autocorrelation between EI and EE. The settling point model might be suitable for long-term control of body weight. Based on the evidence presented, normal-weight subjects are more likely to show signs of biological control and to lose weight when calorie restriction is implemented. It is imperative to discuss the limitations of current research and to suggest better concepts, methods, and studies to improve body weight homeostasis in the future [22].

2. Iron homeostasis and modulation as an example of future studies to treat various diseases like stroke

Redox properties are characteristic of iron. A number of enzymes use it as an active site, and it plays a key role in various cellular and biological functions, such as the production of ATP and DNA. Redox-active iron causes oxidative stress and cell death by generating free radicals and lipid peroxidation. Apoptosis and necrosis are different types of cell death processes, and iron-mediated oxidation plays a central role in ferroptosis. The regulation of iron metabolism and homeostasis is sophisticated. Since hepcidin was identified as the main regulator of iron homeostasis, there have been exciting advances in understanding iron metabolism and regulation. Mammalian cells produce only one iron exporter, ferroportin, which is regulated by hepcidin. Ferroportin is a ferrous iron permease that exports iron to the outside. Recently, iron homeostasis has focused particularly on epigenetics. Iron metabolism is modulated by epigenetic processes including hepcidin. Reviewing recent advances in epigenetic regulation of hepcidin, ferritin, and ferroptosis, we focus on the rapid progress in understanding molecular mechanisms of iron homeostasis. Also discussed is the interaction between iron and methionine oxidation. In addition, several studies have demonstrated that accumulation of brain iron correlates with neuronal damage following stroke. It is briefly discussed how iron metabolism may play a role in strokes. It may be possible to treat various intractable diseases, such as stroke, by understanding the mechanisms behind iron regulation [23].

Author details

Gaffar Sarwar Zaman
Department of Clinical Laboratory Sciences, King Khalid University, Abha, Kingdom of Saudi Arabia

*Address all correspondence to: gffrzaman@gmail.com

References

[1] Gordon Betts J. Anatomy and Physiology. Houston, Texas: OpenStax College. p. 9

[2] Martin E. A Dictionary of Biology. 6th ed. Oxford: Oxford University Press; 2008. pp. 315-316

[3] Biology Online. "Homeostasis". Biology Online. 2019. Archived from the original on 12 August 2020 [Retrieved 27 October 2019]

[4] Kalaany NY, Mangelsdorf DJ. LXRS and FXR: The yin and yang of cholesterol and fat metabolism. Annual Review of Physiology. 2006;**68**:159-191. DOI: 10.1146/annurev.physiol.68.033104.152158

[5] Cannon WB. The Wisdom of the Body. New York: W. W. Norton; 1932. pp. 177-201

[6] Cannon, W. B. (1926). "Physiological regulation of normal states: Some tentative postulates concerning biological homeostatics". In A. Pettit (ed.). A Charles Riches amis, ses collègues, ses élèves (in French). Paris: Les Éditions Médicales. p. 91.

[7] Smith GP. Unacknowledged contributions of Pavlov and Barcroft to Cannon's theory of homeostasis. Appetite. 2008;**51**(3):428-432. DOI: 10.1016/j.appet.2008.07.003

[8] Zorea A. Steroids (Health and Medical Issues Today). Westport, CT: Greenwood Press; 2014. p. 10

[9] Marieb EN, Hoehn KN. Essentials of Human Anatomy & Physiology. 9th ed. San Francisco: Pearson/Benjamin Cummings; 2009

[10] Riggs DS. Control Theory and Physiological Feedback Mechanisms. Baltimore: Williams & Wilkins; 1970

[11] Hall J. Guyton and Hall Textbook of Medical Physiology. (12th ed.). ed. Philadelphia, PA: Saunders/Bich er; 2011. pp. 4-9

[12] Milsum JH. Biological Control Systems Analysis. New York: McGraw-Hill; 1966

[13] Tortora GJ, Anagnostakos NP. Principles of Anatomy and Physiology. Fifth ed. New York: Harper & Row, Publishers; 1987. pp. 315-316, 475, 657-658

[14] Khan Academy. "Homeostasis". Khan Academy. Archived from the original on 20 October 2019 [Retrieved 13 July 2018]

[15] Swedan NG. Women's Sports Medicine and Rehabilitation. USA: Lippincott Williams & Wilkins; 2001. p. 149. Archived from the original on 10 May 2020 [Retrieved 11 October 2019]

[16] Weschler T. Taking Charge of Your Fertility. New York: HarperCollins; 2002. pp. 52, 316, 361-362

[17] West BJ. Where Medicine Went Wrong: Rediscovering the Path to Complexity. Studies of Nonlinear Phenomena in Life Science. Vol. Vol. 11. New Jersey: World Scientific; 2006. DOI: 10.1142/6175. Archived from the original on 6 March 2022 [Retrieved 23 January 2019]

[18] Longo G, Montévil M. Perspectives on Organisms. Germany: Springer; 2014. DOI: 10.1007/978-3-642-35,938-5

[19] Shannon CE, Weaver W. The Mathematical Theory of Communication. 4th print. ed. Urbana: University of Illinois Press; 1963

[20] Rucker R. Mind Tools: The Mathematics of Information.

Harmondsworth: Penguin Books; 1987. pp. 25-30

[21] Koeslag JH, Saunders PT, Wessels JA. The chromogranins and counter-regulatory hormones: Do they make homeostatic sense? Journal of Physiology. 1999;**517**(3):643-649. DOI: 10.1111/j.1469-7793.1999.0643

[22] Müller MJ, Geisler C, Heymsfield SB, Bosy-Westphal A. Recent advances in understanding body weight homeostasis in humans. F1000Res. 2018;**7**:1025. DOI: 10.12688/f1000research.14151.1

[23] Shi H, Almutairi M, Moskovitz J, Xu YG. Recent advances in iron homeostasis and regulation - a focus on epigenetic regulation and stroke. Free Radical Research. 2021;**55**(4):375-383. DOI: 10.1080/10715762.2020.1867314

Section 2

Recent Advances

Chapter 2

Characterization of Homeostatic Level Based on Non-Linear Variables of Heart Rate Variability

Moacir Fernandes de Godoy and Michele Lima Gregório

Abstract

Heart Rate Variability (HRV) has been frequently cited as an indicator of homeostatic status. Low levels of HRV are associated with aging, disease, or increased risk of death. The authors based this chapter on an alphanumerical classification for the levels of homeostasis, structured on three linear variables (Heart Rate, RMSSD, and HF ms2) by analyzing a bigdata with more than 30 million pieces of data collected from literature. It was possible to confirm the clinical validity of this alphanumeric classification. It has been mentioned that HRV analysis in time and frequency domains are often not sufficient to characterize the complex dynamics of the heartbeat. Thus, the primary objective of this study was to verify whether or not there are correlations between the variables of the non-linear domain with variables and indices of the linear domain and also with the homeostatic level of individuals. It was found, contrary to expectations, that the variables ApEn, SampEntr and DFA $\alpha 1$ were not useful in characterizing the homeostatic level, since they do not differentiate between healthy and highly compromised individuals. Regarding the parasympathetic, sympathetic and stress indexes, only DFA a1 detected a correlation with the sympathetic index and the stress index.

Keywords: heart rate variability, non-linear, autonomic nervous system, homeostasis, big data

1. Introduction

Heart rate variability has been frequently cited as a relevant indicator of homeostatic status, since low levels of heart rate variability, especially those related to the parasympathetic component, are repeatedly associated with the presence of aging, disease, or increased risk of death [1].

Generally, the evaluated variables are categorized as belonging to the time domain, frequency domain and non-linear domain. Among the linear variables with clinical importance, most mentioned in the literature, heart rate, RMSSD (root mean square of successive differences between normal heartbeats) and HFms2 (high-frequency component), stand out.

Both the time domain and frequency domain methods assume that HRV signals are linear, and thus cannot quantify the dynamic structure of the signal.

More recently, there has been an increase in references suggesting that variables in the non-linear domain of HRV would be more sensitive in detecting alterations in homeostasis.

Sassi et al., in 2015 [2] evaluated the available literature over a period of 18 years (1996 to 2013) and found only 21 studies with more than 200 cases, in which non-linear HRV assessment methods were used.

In addition, a lot of discrepancy between studies has been observed, calling attention to the lack of standardization and proper validation of some methods used [3].

To assess the non-linear properties, several methods have been proposed in the past, including Fractal Dimension, Lyapunov Exponent, Hurst Exponent, Correlation Dimension, Approximate Entropy, Sample Entropy, Shannon Entropy and Detrended Fluctuation Analysis. All these methods quantify some non-linear property of HRV [3]. Among these various HRV variables belonging to the non-linear domain, three stand out in the world Literature, namely Approximate Entropy (ApEn), Sampling Entropy (SampEnt) and the alpha 1 component of the Detrended Fluctuation Analysis (DFA $\alpha 1$).

ApEn is a statistic quantifying regularity and complexity of a stationary signal [4]. This means that ApEn quantifies the predictability of fluctuations in the time series. The main idea behind approximate entropy is that a sequence is regular if a subsequence and an expansion of the subsequence are similar [5].

Byum et al. in 2019 [6], studied a total of 33 patients with major depressive disorder (MDD) based on the DSM-IV criteria, and 33 healthy controls, matched for age and gender. Four entropy indices, approximate entropy, sample entropy, fuzzy entropy and Shannon entropy, were extracted. There were no significant differences in entropy features between the control and patient groups at the base line. The authors considered that this inconsistency was likely a result of the heterogeneous presentation and multifactorial etiology of MDD.

Garner et al., in 2021 [7] examined 38 subjects with Chronic Obstructive Pulmonary Disease (COPD) and 38 matched controls. They measured heart rate variability, through Approximate Entropy (ApEn), during 30 minutes, in the supine position, without any physical, sensory or pharmacological stimuli. They concluded that ApEn was capable of optimally identifying the decrease in chaotic response in COPD but, despite this, ApEn should be considered a relatively unpredictable mathematical marker and the use of other techniques to evaluate a healthy or pathological condition needs to be encouraged.

Beckers et al., [3], evaluated the Approximate Entropy behavior in 21 patients in advanced stages of heart failure (NYHA class III and IV) compared to 21 healthy individuals, age and sex matched. Twenty-one heart failure (CHF) patients (NYHA class III and IV; all males; age: 54.5 ± 2.9 years) were included. An age and sex matched group of 21 healthy subjects (all males; age: 54.5 ± 4.1 years) was used as a control population. No statistically significant differences were found between groups regarding Approximate Entropy.

Sample Entropy (SampEn) has been proposed as a method to overcome limitations associated with approximate entropy (ApEn). Para Aboy et al., [8], SampEn is more consistent and agrees more closely with theory for known random processes than ApEn.

Al-Angari and Sahakian in 2008 [9] used Sample Entropy, as a measure of signal complexity to evaluate the behavior of heart rate variability in Obstructive Sleep Apnea Syndrome (OSAS). They found that healthy subjects have significantly more complex HRV pattern than the OSA subjects and that the sample entropy had an accuracy of 70.3%. They stressed, however, that the sample entropy approach does not show major improvement over the existing methods. In fact, its accuracy in detecting sleep apnea was relatively low in those well classified patients.

The alpha 1 exponent of the Detrended Fluctuation Analysis (DFA α1) is a non-linear domain variable in the analysis of heart rate variability and represents the observed self-similarity or fractal nature between the RR intervals in a time series. Tapanainen et al., in 2002 [10] showed the evolution of 697 survivors after acute myocardial infarction, through the analysis of conventional variables in the time and frequency domains, in addition to the non-linear variable DFA α1. They found that (49) 7% of cases died after a mean follow-up of 18.4 ± 6.5 months. The DFA alpha 1 exponent with a value below 0.65 was an independent predictor of death, both in univariate as in multivariate analysis, regardless of the presence or absence of left ventricular dysfunction.

Schaffarczyk et al., in 2022 [11] evaluated the usefulness of DFA α1 with cutoff values of 0.75 and 0.50 in determining aerobic and anaerobic thresholds, in 26 female volunteers aged between 20 and 59 years, having found a good correlation of values according to the intensity of the exercise. At low exercise intensity, DFA α1 values indicated a well-correlated fractal pattern remaining close to 1.0 or slightly above. As the effort intensity was increased, they noticed a decrease in the index to approximately 0.75, approaching the uncorrelated random patterns, represented by values close to 0.50 or even below, in the case of even more intense work. It was inferred that this index may reflect the state of systemic internal load.

These sometimes conflicting observations, highlight the importance of the issue addressed here, which aims to characterize and quantify the real practical use of the most widespread and clinically applicable non-linear methods for assessing HRV.

2. Method

2.1 Validation

In 2022, Godoy and Gregório [1] evaluated the Heart Rate Variability as a Marker of Homeostatic Level, proposing an alphanumeric classification, after collecting about 10.5 million data from a bigdata, based on specific changes in three variables of the linear domain of the HRV. At that moment 465,966 data from those variables were detected, 387,638 related to heart rate, 45,545 related to RMSSD and 32,783 related to HFms2.

A total healthy individual, with an excellent Homeostatic Level and, therefore, with very low risk, would receive the A1B1C1 classification. An individual with a high basal heart rate, a very low RMSSD and HF power values would be classified as A3B3C3 indicating high severity, low homeostatic level and, therefore, at high risk. Several intermediate combinations would be possible characterizing the current state of each case. The cutoff values for each variable were:

Level A: Heart Rate (bpm)
- Stage Al: Heart Rate less than 70 bpm
- Stage A2: Heart Rate between 70 and 85 bpm
- Stage A3: Heart Rate above 85 bpm

Level B: RMSSD (ms)
- Stage B1: RMSSD above 32 milliseconds
- Stage B2: RMSSD between 32 and 28 milliseconds
- Stage B3: RMSSD less than 28 milliseconds

Level C: HF ms2
- Stage C1: HF ms2 above 468 ms2

Stage C2: HF ms2 between 468 and 156 ms2
Stage C3: HF ms2 less than 156 ms2

In the present study, we initially sought to validate these data based on an update of available big data. Our database now has more than 31.6 million pieces of data, of which 10,671,458 were related to the three classic linear variables (554,483 to Heart Rate; 5,176,138 to RMSSD; 4,940,837 to HFms2) and 30,723 to the three non-linear chosen variables (6670 to ApEn, 6785 to SampEntr and 17,268) from DFA α1. The complementary amount of data refers to other variables not included in this study. The references that enabled the construction of this bigdata can be made available upon request.

The variables analyzed in the present study and components of the bigdata were grouped according to the clinical status of the participating individuals as belonging to the group of healthy individuals and the group with significant impairment of the homeostatic level, such as, for example, cases of neoplasms, liver dysfunctions or advanced kidney disease, hospitalizations in intensive care units in critical situations, prematurity and conditions of imminent death. **Figure 1** presents the distribution of the quantities of analyzed variables in these two situations.

It can be seen that, the cut-off levels proposed with the initial sample remained equivalent in the current magnified sample and, therefore, become validated.

2.2 Study sample

Based on the proposed alphanumeric code, an unselected group of 123 individuals from the personal casuistic was evaluated, regardless of sex and age, with different clinical states, from apparently healthy to severely compromised health status. **Figure 2** specifies the classification of cases in each impairment group, either at level I, involving patients with three stages 1 or two stages 1 or without duplication of stages; at level II, involving patients with three stages 2 or two stages 2; at level III, involving patients with three stages 3 or two stages 3.

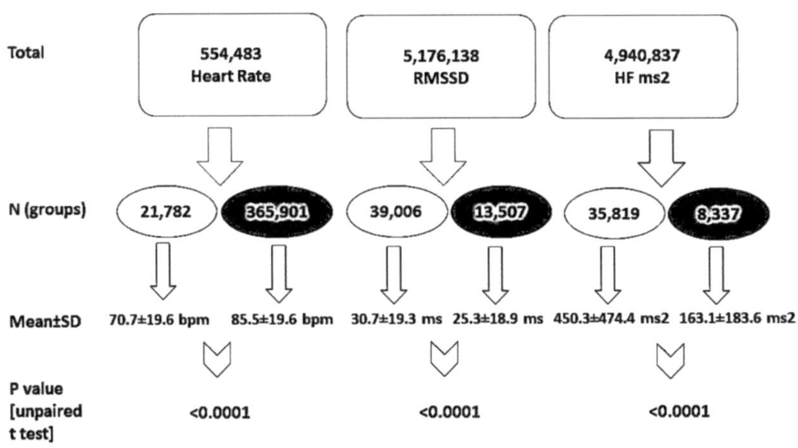

Figure 1.
Data distribution, from the bigdata of 31.6 million pieces of data, with segmented relation to the three selected variables, and according to the group of healthy individuals (light sets) and the group with severe homeostatic impairment (dark sets). The complementary amount of data refers to variables not included in this study. Variable values are presented as mean ± standard deviation. P values are relative to intergroup comparisons of homeostatic impairment.

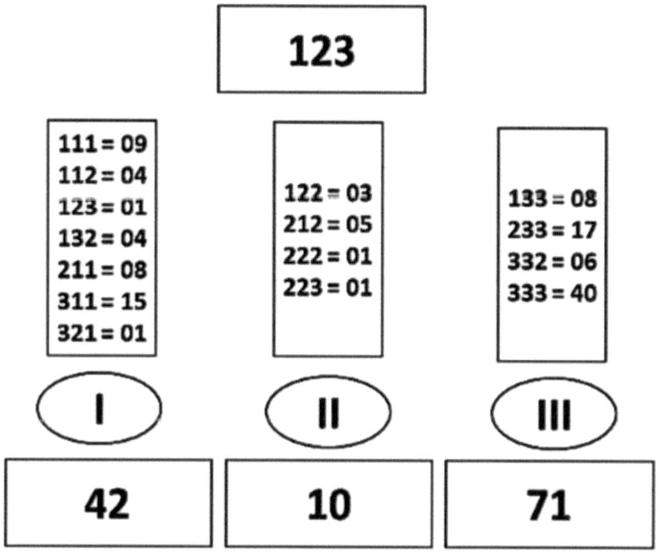

Figure 2.
Distribution of homeostatic level classifications of 123 non selected patients, based on the alphanumeric coding proposed by Godoy and Gregório (2022) [1].

2.3 Statistical analysis

Continuous quantitative data were analyzed using the unpaired Student's t-test. Discrete or non-Gaussian quantitative data were analyzed using the Mann-Whitney test. For correlations, both Pearson's correlation and Spearman's correlation were applied, depending on the type of variable. The graphical representation was made using Boxplot and linear regression graphs. An alpha error of 5% was admitted, with P values lower than or equal to 5% being considered significant. The statistical software used was StatsDirect version 3.3.5. To quantify the HRV variables, the Kubios HRV Scientific application version 4.0.1 of October 2022 was used.

3. Results

We sought to assess whether each of the three non-linear variables (ApEn, SampEn and DFA alpha 1) were correlated or not with the level of homeostatic impairment. It was observed that none of the evaluations detected a significant correlation between the variable and the homeostatic level (Spearman's rank correlation coefficient); **Figure 3(a–c)**.

Comparisons were also made of each of the three non-linear variables between individuals considered healthy (Group I, light sets) and those with severe health impairment (Group III, dark sets). It was possible to collect a total of 30,723 pieces of information on these three variables. It was found that none of the three non-linear variables studied was able to discriminate these opposing groups of homeostasis impairment (**Figures 4**–7).

Finally, Pearson's correlations were sought between the non-linear and the linear variables, between the non-linear and the parasympathetic, sympathetic and stress

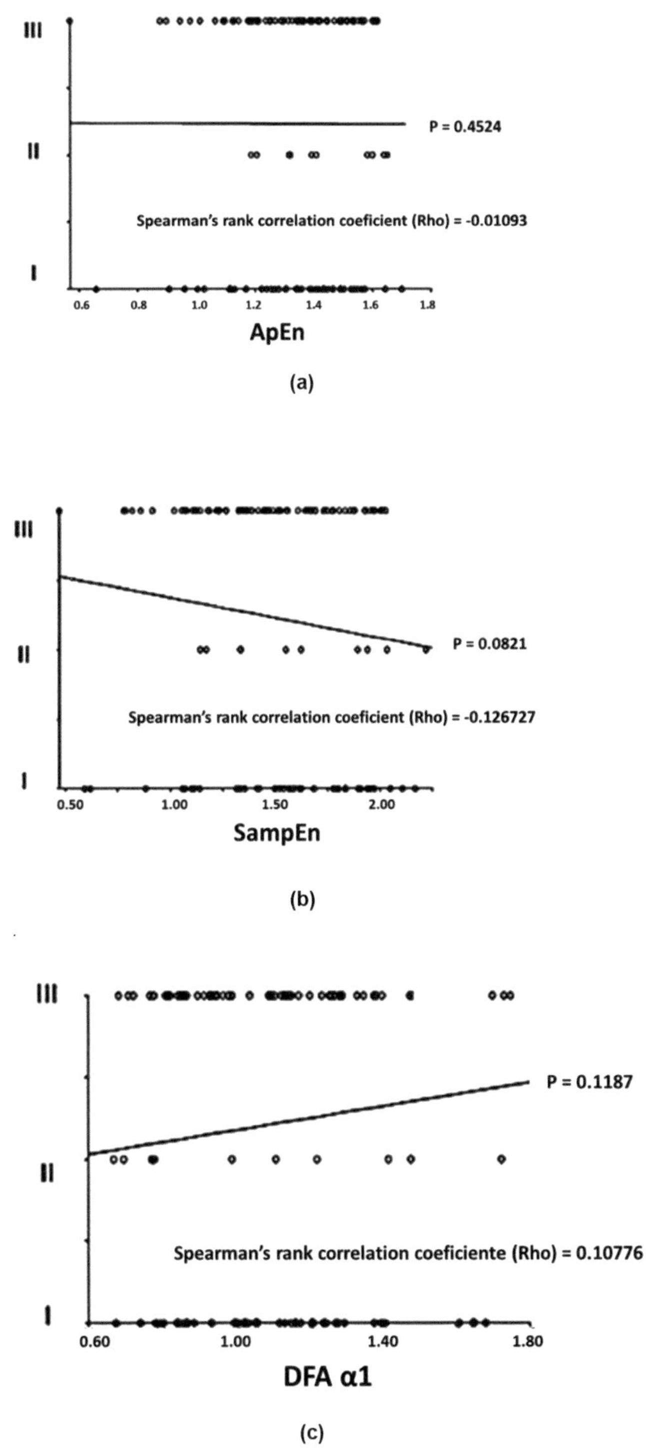

Figure 3.
a. Spearman's rank correlation coefficient between the non-linear variable ApEn and the homeostatic levels. b. Spearman's rank correlation coefficient between the non-linear variable SampEn and the homeostatic levels. c. Spearman's rank correlation coefficient between the non-linear variable DFA α1 and the homeostatic levels.

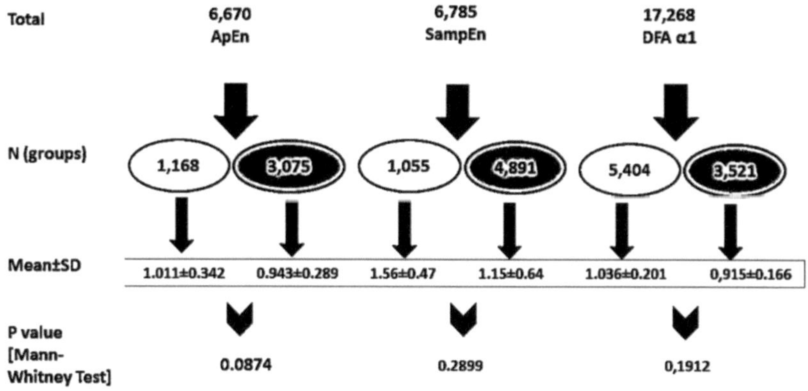

Figure 4.
Statistical comparison by the non-parametric Mann-Whitney test on the ability of the non-linear variables, ApEn, SampEn and DFA alpha 1 to distinguish between healthy individuals and individuals with significant homeostatic impairment.

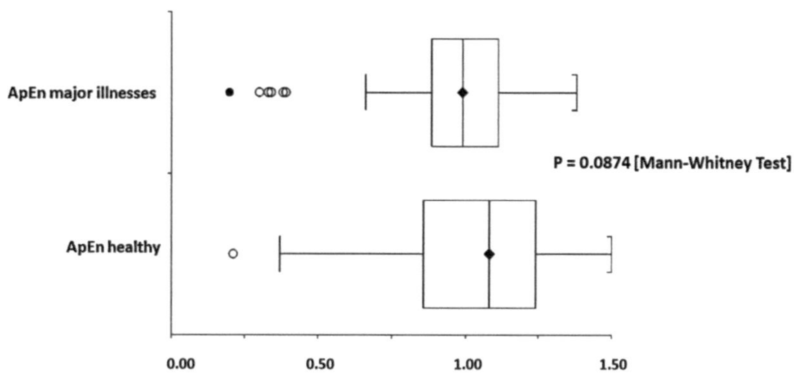

Figure 5.
Boxplot comparing the values of the non-linear variable ApEn, in healthy individuals and in those with significant homeostatic impairment.

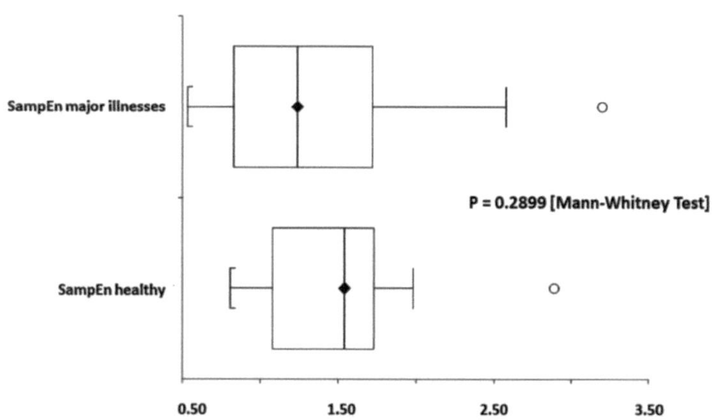

Figure 6.
Boxplot comparing the values of the non-linear variable SampEn, in healthy individuals and in those with significant homeostatic impairment.

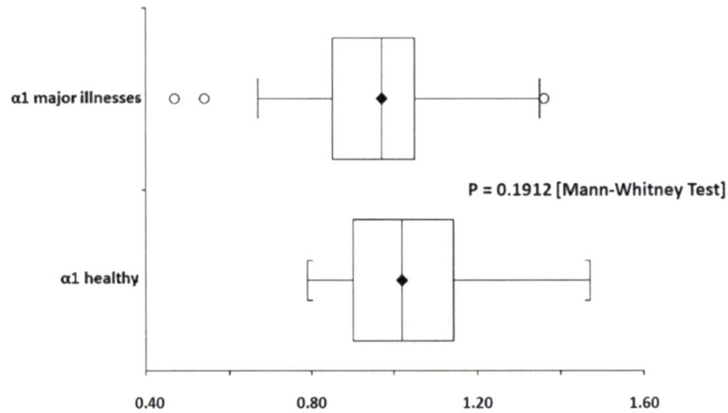

Figure 7.
Boxplot comparing the values of the non-linear variable DFAα1, in healthy individuals and in those with significant homeostatic impairment.

	Heart Rate	RMSSD	HFms2	PNS	SNS	Stress Index	Groups
ApEn	0.0429	−0.0794	−0.0637	0.3385	−0.2474	−0.1763	−0.0025
Samp. Entr	−0.0568	−0.0949	−0.0983	0.3280	−0.2241	−0.1392	−0.1135
DFAα1	−0.0819	−0.1278	−0.1157	0.1729	−0.4869*	−0.5745*	0.1107

* = *significant correlation.*

Table 1.
Pearson correlation (r values) between non-linear variables and linear variables, autonomic indexes and clinical conditions.

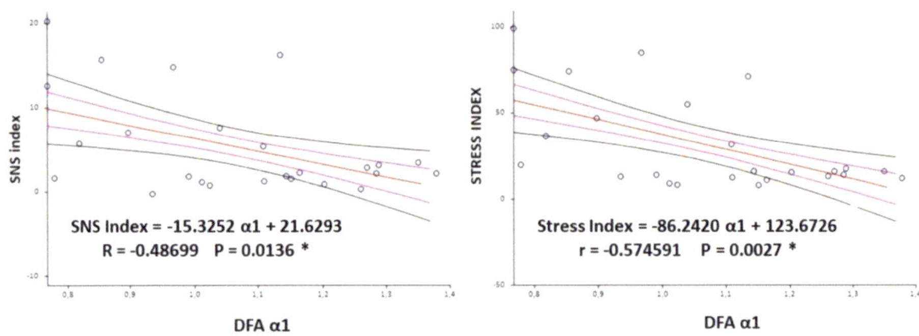

Figure 8.
Linear correlations between the non-linear variable DFA α1 and the sympathetic and stress indexes [= significant].*

indexes and also between the non-linear variables and the clinical classification of health states.

It is worth remembering that the PNS index (parasympathetic nervous system tone index) is derived from the association between the variables Mean RR, RMSSD and SD1 indicates a parasympathetic nervous system activity compared to normal resting values. The SNS index (Sympathetic nervous system tone index) is derived from the association between the variables Mean HR, Stress index and SD2 indicates Sympathetic nervous system activity compared to normal resting values. The Stress index is the square root of the Baevsky's stress index with normal values ranging from 7 to 12.

In general, there were no significant correlations between the analyzed variables, that is, the correlation coefficients were not significantly different from zero, except for the negative correlations between DFA a1 and the sympathetic index (r = −0.48699; P = 0.0136) and DFA α1 with the Stress Index (r = −0.574591; P = 0.0027) (**Table 1** and **Figure 8**).

4. Conclusions

I. It was possible to confirm the clinical validity of an alphanumeric classification of homeostatic level, based on a big data with more than 30 million data related to variables of Heart Rate Variability, collected in the world Literature, by comparing the mean values obtained with the mean expected values.

II. The three classical variables of the linear domain (Heart Rate, RMSSD and HFms2), based on the suggested cut-off levels, are significantly different when comparing healthy individuals and individuals with significant impairment of homeostasis.

III. No correlation was detected, using Spearman's rank correlation coefficient, between each of the three non-linear variables (ApEn, SampEn and DFA α1) and the degree of homeostatic impairment.

IV. There are no correlations between the variables of the non-linear domain with variables of the linear domain, neither with the degree of global clinical impairment nor homeostatic level of individuals.

V. The three variables of the selected non-linear domain (ApEn, SampEn and DFA $\alpha 1$), are not statistically different when healthy individuals are compared with individuals with significant impairment of the homeostatic level.

VI. In general, there were no significant correlation by Pearson's correlation coefficient between the non-linear variables and the selected linear variables, as well as between the non-linear variables and the parasympathetic index. Only DFA $\alpha 1$ showed a significant correlation (negative and moderate) with the sympathetic index (r = – 0.48699; P = 0.0136) and with the Stress Index (r = −0.574591; P = 0.0027).

VII. These findings suggest that the non-linear variable DFA $\alpha 1$ may be considered a marker of individual stress burden.

VIII. Finally, the relevance of the non-linear evaluation was demystified since the non-linear variables evaluated did not show significant discriminatory power.

Acknowledgements

This work has been supported by the Brazilian research agency CNPq. The first author was funded by the grant 313009/2021-0 (Research Productivity Scholarship).

Author details

Moacir Fernandes de Godoy[1,2]* and Michele Lima Gregório[1,3]

1 Sao Jose do Rio Preto Medical School - Famerp, SP, Brazil

2 Transdisciplinary Nucleus for Chaos and Complexity Studies, SP, Brazil

3 Cedars-Sinai Medical Center, CA, United States

*Address all correspondence to: mf60204@gmail.com

References

[1] de Godoy MF. Nonlinear analysis of heart rate variability: A comprehensive review. Journal of Cardiology and Therapy. 2016;**3**(3):528 533. Availablc from: http://www.ghrnet.org/index.php/jct/article/view/1724

[2] Sassi R, Cerutti S, Lombardi F, Malik M, Huikuri HV, Peng C-K, et al. Advances in heart rate variability signal analysis: Joint position statement by the e-Cardiology ESC Working Group and the European Heart Rhythm Association co-endorsed by the Asia Pacific Heart Rhythm Society. Europace. Sep 2015;**17**(9):1341-1353. DOI: 10.1093/europace/euvo15

[3] Beckers F, Ramaekers D, Aubert AE. Approximate entropy of heart rate variability: Validation of methods and application in heart failure. Cardiovascular Engineering. 2001;**1**:177-182. DOI: 10.1023/A:1015212328405

[4] Pimcus SM. Approximate entropy as a measure of system complexity. Proceedings of the National Academy of Sciences of the United States of America. 1991;**88**:2297-2301

[5] Mayer CC, Bachler M, Hõrtenhuber M, Stockers C, Holzinger A, Wassertheurer S. Selectfon of entropymeasure parameters for knowledge discovery in heart rate variability data. Bioinformatics. 2014;**15**(Suppl. 6):S2. Available from: http://www.biomedcentral.com/1471-2105/15/S6/S2

[6] Byun S, Kiim AY, Jang EH, Kiim S, Choi KW, Yu HY, et al. Entropy analysis of heart rate variability and its application to recognize major depressive disorder: A pilot study. Technology and Health Care. 2019;**27**(81):407-424. DOI: 10.3233/THC-199037

[7] Garner DM, Bernardo AFB, Vanderlei LCM. HRV analysis: Unpredictability of approximate entropy in chronic obstructive pulmonary disease. Series of Cardiology Research. 2021;**3**(1):1-10

[8] Aboy M, Cuesta-Frau D, Austin D, & Mico-Tormos P. Characterization of sample entropy in the context of biomedical signal analysis. In: 2007 29th Annual International Conference of the IEEE Engineering in Medicine and Biology Society. 2007. DOI: 10.1109/iembs.2007.4353701

[9] Al-Angari HM, Sahakian AV. Use of sample entropy approach to study heart rate variability in obstructive sleep Apnea syndrome. IEEE Transactions on Biomedical Engineering. Oct 2007;**54**(10):1900-1904. DOI: 10.1109/TBME.2006.889772

[10] Tapanainen JM, Thomsen PEB, Køber L, Torp-Pedersen C, Makikallio TH, Still AM, et al. Fractal analysis of heart rate variability and mortality after an acute myocardial infarction. The American Journal of Cardiology. 2002;**90**(4):347-352

[11] Schaffarczyk M, Rogers B, Reer R, Gronwald T. Validation of a non-linear index of heart rate variability to determine aerobic and anaerobic thresholds during incremental cycling exercise in women. European Journal of Applied Physiology. 2022;**123**(2):299-309. DOI: 10.1007/s00421-022-05050-x

Chapter 3

Homeostatic Control of Neuronal Activity

Cato Drion

Abstract

For healthy brain functioning, it is crucial that neuronal networks do not become hyperactive, but also, that they remain excitable. Homeostatic mechanisms ensure that neuronal activity remains within a functional range. How does that work? In this chapter, we will explore homeostatic control of neuronal activity. We will start by introducing the basics of neuronal communication to establish what makes a neuron excitable. Then, we will learn how neurons are able to tune their own excitability, which is called *homeostatic intrinsic plasticity*. Next, we will discuss the ability of neurons to tune the strength of their connections to other neurons. This is called homeostatic *synaptic* plasticity and involves synaptic scaling, the up- and downregulation of receptors, and the control of neurotransmitter release. Finally, we will review the role of glia in neuronal network homeostasis and discuss disorders where the homeostatic control of neuronal activity is compromised.

Keywords: neuronal excitability, homeostatic intrinsic plasticity, homeostatic synaptic plasticity, synaptic scaling, neurotransmitter release, receptors

1. Introduction

All our thoughts, feelings, and actions are enabled by the cells of our nervous system: neurons. Neurons communicate with each other and with the rest of our body, our organs, senses, and muscles. Neuronal activity is essential to function, but at the same time, neuronal activity should not become excessive. To understand how neuronal activity is under homeostatic control, we first need to understand the basics of neuronal activity.

1.1 The electrochemical basics of neuronal activity

Neurons communicate via electrochemical signals. They pass on currents to one another through the extracellular space and across the neuronal cell membrane. These currents are carried by ions, primarily sodium (Na^+), potassium (K^+), and chloride (Cl^-), that can pass the neuronal membrane through specialized pores: ion channels. Neurons can communicate this way because they are electrically charged: there is a difference in charge between the extracellular and intracellular space, called the membrane potential. At rest, the membrane potential of a neuron is about –70 millivolts (mV). From this baseline value, the neuron can either depolarize (the membrane potential becomes less negative) or hyperpolarize (the membrane potential becomes even more negative), depending on what messages the neuron receives. To be able to receive and respond to these changes in membrane potential, a neuron needs to maintain a stable resting membrane potential.

The membrane potential depends on the distribution of ions across the neuronal membrane. Therefore, it is crucial that the concentrations of ions in the intracellular and extracellular space are tightly regulated. Part of this is done by transporters that actively carry ions in and out of the cell. The most important transporter continuously exchanges three Na^+ ions for two K^+ ions and is therefore called the sodium-potassium pump. But also, glial cells play an important role herein. Astrocytes, for instance, buffer K^+ ions and ensure that ions are distributed across neuronal networks. These examples illustrate how neurons maintain a stable baseline, when they are at rest. But what happens during neuronal activity?

If an incoming signal from the environment (another neuron, for example) triggers a strong enough depolarization, a neuron can generate action potentials, which are the "messages" neurons send to one another. An action potential is a short, transient, and high-amplitude depolarization of the membrane potential, which only occurs if the membrane potential surpasses a threshold value. This has everything to do with the type of ion channels involved: voltage-gated ion channels.

Voltage-gated ion channels only open when the membrane potential is between specific values. They also close at specific values, creating a limited window for ions to cross the membrane. This ensures that the membrane potential can be restored to baseline values after neuronal activity. In addition, the currents carried by specific ions also behave in a voltage-dependent way. In general, Na^+ will only flow into a neuron (thereby generating a positive inward current that depolarizes the membrane potential) if the membrane potential is negative. K^+, on the other hand, is more inclined to leave the neuron (thereby generating a positive outward current that hyperpolarizes the membrane potential) if the membrane potential is positive. How much of a particular ion can cross the membrane, and thus, how much current it can generate, depends on the amount of *open* ion channels. This is then dependent on the membrane potential, but also on the *amount* of ion channels available. As we will see in this chapter, neurons can regulate the amount of *voltage-dependent* ion channels available in the membrane. To summarize, the membrane potential, the distribution of ions, and the amount of voltage-gated ion channels all determine whether or not a neuron will generate an action potential (**Table 1**).

How well a neuron is capable of generating an action potential reflects its *excitability*. It is crucial that neuronal excitability remains within limits; it should not be too hard, but also not too easy, for a neuron to *fire* an action potential. Several mechanisms contribute to keeping neuronal excitability within limits, as we will see later on. First, we will discuss how an action potential is transmitted from neuron to neuron—in other words: how do neurons talk to each other?

1.2 The synapse

The site where neurons connect, and electrochemical signals are being transmitted from one neuron to another is called the synapse (**Figure 1**). In the most common type of synapse, the transmission happens from the *axon terminals* of the presynaptic ("sending") neuron to the dendritic spines of the postsynaptic ("receiving") neuron. Transmission of the electrochemical signals across the synapse usually requires the presynaptic neuron to release a neurotransmitter at its axon terminal. A neurotransmitter is a

Term	Description
Ions	Charged particles: sodium (Na^+), potassium (K^+), chloride (Cl^-) and calcium (Ca^{2+}) are important
Membrane potential	The difference in charge between the intra- and extracellular space
Action potential	A rapid, high-amplitude depolarization of the membrane potential that can be passed on to neighboring neurons and thus serves as a "message"
Neuronal excitability	How easily can a neuron be prompted to generate an action potential?
Voltage-gated ion channels	Channels that open, close, and inactivate in response to changes in the membrane potential

Table 1.
Important definitions in neuronal activity.

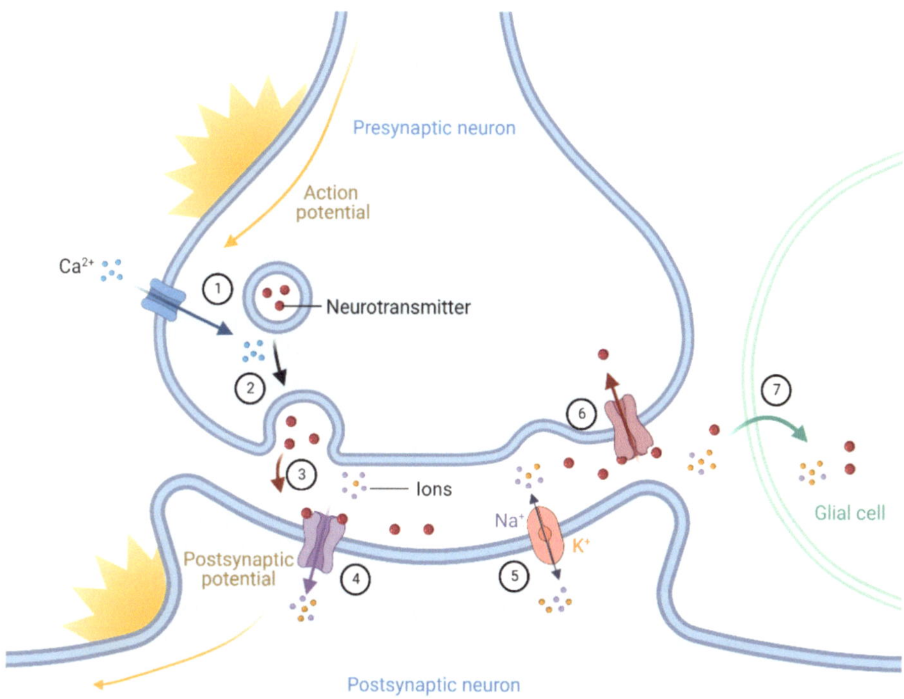

Figure 1.
(created with BioRender.com): the biochemical basis of neuronal communication. A simplified drawing of the typical chemical synapse. (1) In the presynaptic neuron, an action potential arrives at the axon terminal, where it triggers calcium (Ca^{2+}) influx. (2) Ca^{2+} triggers the fusion of neurotransmitter-containing vesicles with the neuronal membrane and consequently, neurotransmitter is released into the synaptic cleft. (3) The neurotransmitter binds to the receptors on the dendritic spine of the postsynaptic neuron. This will—either directly or indirectly—cause ions to flow across the postsynaptic membrane (4) and thereby change the membrane potential (4) —a postsynaptic potential (which can be either inhibitory or excitatory) is created. Following neuronal activity, a number of processes help to restore activity levels. (5) The sodium-potassium pump exchanges three sodium (Na+) ions for two potassium (K+) ions to restore or maintain constant ion distributions. Neurotransmitters are degraded, taken up by the presynaptic neuron or (6) bind to presynaptic autoreceptors that regulate neurotransmitter release. (7) Glia take up neurotransmitters and ions.

signaling molecule that usually binds as a *ligand* for a specific receptor. Neurotransmitter release is initiated by a signal from another important ion: calcium (Ca^{2+}). While the other cations (Na^+, K^+) are the most important players in the membrane potential and the action potential, Ca^{2+} acts as a signaling molecule in all kinds of intracellular chain reactions. In the presynaptic axon terminal, the chain reaction starts when an action potential arrives there (the presynaptic neuron "fires"). The action potential depolarizes the membrane potential, which opens voltage-gated Ca^{2+} channels. This triggers Ca^{2+} influx into the axon terminal, which in turn causes neurotransmitter-containing vesicles to fuse with the presynaptic membrane and release the neurotransmitter into the synaptic cleft. There, neurotransmitter molecules bind with their receptors on the postsynaptic neuron. The receptors then get activated and either directly or via second messengers, allow ion channels to open, which causes ions to flow through and change the postsynaptic membrane potential: the electrochemical message has been received.

Remember that the ion channels in this description are (either directly or indirectly) operated by receptors that are activated by neurotransmitters. We call those channels "ligand-gated" ion channels (see **Figure 1**).

Excitatory neurons send an activating message: they release neurotransmitters that induce a depolarization of the postsynaptic neuron. The most important excitatory neurotransmitter is glutamate. Glutamate is a ligand for glutamatergic receptors such as the AMPA receptor (AMPA-R) and the NMDA receptor. Throughout this chapter, we will repeatedly encounter the AMPA-R in homeostatic processes. The AMPA-R is a ligand-gated ion channel: when glutamate binds to the AMPA receptor, the conformation of the receptor changes and this allows Na^+ to enter the neuron. The entry of Na^+ depolarizes the membrane potential, we call this an excitatory postsynaptic potential (EPSP).

Inhibitory neurons release neurotransmitters that result in a hyperpolarization of the membrane potential of the postsynaptic neuron; they generate inhibitory postsynaptic potentials (IPSPs). For example, their receptors will allow anions (negative ions, here: Cl^-) to enter the neuron, or allow potassium (K^+) ions to leave the cell. Both mechanisms hyperpolarize the cell, making it more difficult to generate an action potential. **Table 2** lists the most important terms in synaptic transmission.

Neurons make many connections with neighboring neurons: they form neuronal networks. A neuron can have thousands of synapses, and hence, receives synaptic input from a large number of excitatory and inhibitory numbers, in other words: it gets messages to tell it to fire, but at the same time, surrounding neurons may send messages to prevent the firing. For neuronal functioning, it is important that the balance between excitatory input and inhibitory input is maintained, within the neuron (the amount of inhibitory and excitatory synapses), but also at the network level (the amount of inhibitory and excitatory cells in a network). We will see that this balance between excitation and inhibition is under homeostatic control. To understand how this happens, we must first introduce neuronal plasticity.

1.3 Neuronal plasticity

Neuronal networks are not static—in fact, the connections within a neuronal network change all the time. This high degree of plasticity has been primarily demonstrated in brain areas associated with learning and memory, and indeed, it seems to underlie our ability to learn.

In 1949, the Canadian psychologist Donald Hebb proposed that if one neuron repeatedly fires together with another neuron, the connection between those neurons will be reinforced. The phrase "neurons that fire together, wire together" is

Term	Description
Synapse	Contact point between neurons, where signals are sent presynaptically and received postsynaptically
Axon	Protrusion from the neuron that conducts signals from the cell body to the synapse ("sending")
Dendrite	Protrusion from the neuron that conducts signals from the synapse to the cell body ("receiving")
Ligand-gated ion channel	Ion channel that is activated when a ligand binds to it, or when a ligand binds to the receptor that indirectly activates it
Neurotransmitter	Signaling molecule that enables synaptic signaling by binding to the receptor for which it is the ligand
Receptor	Protein that binds neurotransmitters, which then triggers a response in the receiving neuron

Table 2.
Important definitions in synaptic transmission.

often used to illustrate this structural plasticity that we still call Hebbian Learning. Hebbian learning is a form of structural synaptic plasticity: the shaping of connections between neurons occurs by strengthening or weakening specific synapses. We will briefly illustrate how this works, so we can compare and contrast the underlying mechanisms to homeostatic synaptic plasticity later on.

Repeated simultaneous activation of a pre- and postsynaptic neuron will trigger the influx of Ca^{2+}, not only at the pre-synapse (remember how Ca^{2+} influx initiates presynaptic neurotransmitter?) but also in the postsynaptic cell. Here, Ca^{2+} can enter the cell through voltage-gated Ca^{2+} channels and NDMA receptors, but Ca^{2+} is also released from intracellular stores in response to stimulation. The activity-dependent rise in intracellular Ca^{2+} triggers specialized proteins to take action: they are Ca^{2+}-dependent kinases (CaMK) that phosphorylate other proteins and thereby trigger intracellular processes. In this case, consider CaMKII: this kinase phosphorylates subunits of the AMPA receptors, making those more active. Furthermore, Ca^{2+} will set off signaling cascades resulting in the synthesis of proteins that enhance synaptic transmission, for example, extra receptors, or the building blocks of those receptors (subunits). Or, proteins are produced that transport additional receptors and insert them into the membrane, or proteins that keep the receptors "trapped" in the membrane (these are called postsynaptic density proteins). In addition to protein synthesis, the signaling cascades can also stimulate the formation of contact points (postsynaptic dendritic spines) between neurons.

These processes collectively make the postsynaptic neuron more responsive and thus, strengthen the connection at that specific synapse. The next time this synapse gets stimulated, its response will be stronger. This is a long-lasting effect (up to months in the rodent hippocampus) and hence we call this long-term potentiation (LTP) [1]. The counterpart of LTP is called long-term depression (LTD), which happens at synapses that are not used enough: these connections are weakened or even removed. Hebbian learning occurs through enhancing the relevant connections with LTP and removing the irrelevant ones through LTD, thus shaping the brain through experience. We will later refer to this as "Hebbian plasticity." Of note, the example mechanisms described here are not all there is to it. For instance, you may have noticed that we have discussed post-synaptic LTP mechanisms only, but there are also—albeit less (known) mechanisms taking place in the pre-synapse.

Now you know the basics of neuronal activity and neuronal plasticity, we will dive into homeostatic control of neuronal activity.

2. Homeostatic control of neuronal activity

In the brain, there is a high degree of plasticity. Neuronal networks shape their connections according to experience, and this underlies our ability to learn. Neurons are able to selectively strengthen and weaken their synapses, which reinforces their excitability. However, neurons should not become too easy to excite: hyperexcitable neurons lose the ability to respond selectively to relevant inputs, which disables normal functioning. Furthermore, as we will see later on, the hyperexcitability of neuronal networks can lead to epilepsy. But it should also not become too hard to excite a neuron: without responsive neurons, neuronal networks would fall silent. Thus, the excitability of neurons and neuronal networks needs to be kept within functional limits. In the healthy brain, neurons compensate for hyperactivity or lack of activity using several mechanisms. These mechanisms aim to restore excitability to baseline levels and keep activity of neuronal networks within limits. In other words, these mechanisms constitute homeostatic control of neuronal activity.

In this chapter, we will explore the mechanisms of different types of homeostatic plasticity. Homeostatic plasticity can be divided into two categories: homeostatic intrinsic plasticity and homeostatic synaptic plasticity. The latter pertains to mechanisms influencing connections between neurons, mostly targeting the functioning of ligand-gated ion channels and neurotransmitters. Homeostatic intrinsic plasticity concerns a neuron's own baseline level of excitability. This type of homeostatic plasticity mostly concerns voltage-gated ion channels, as we will discover in the next section.

2.1 Homeostatic intrinsic plasticity

Different types of neurons usually have their own typical pattern of activity: specific characteristics that are called intrinsic firing properties. Intrinsic firing properties are genetically imprinted during the early stages of neurodevelopment, by the expression of specific transcription factors. Examples of these intrinsic firing properties are: how difficult or easy it is to generate an action potential (the threshold), how often the neuron can generate consecutive action potentials (the firing frequency range), what currents specifically play a role in the action of this neuron (depending on what type of ion channels the neuron expresses) et cetera. Intrinsic firing properties dictate the typical behavior of a neuron and are used to discriminate between types of neurons.

Interestingly, the fact that these intrinsic firing properties are fixed does not mean that there is no room for plasticity: rather than one value for each property, intrinsic firing properties are defined as a range, and within that range a neuron can adapt its behavior. This means that within certain limits, a neuron can use cellular mechanisms to adjust its excitability. This is what happens in homeostatic intrinsic plasticity.

Homeostatic intrinsic plasticity is simply put: neurons sensing their own excitability and consequently adjusting it in a compensatory manner. For example, if a neuron has been inactive for a while, it will increase its responsiveness to subsequent input, as if to become more sensitively tuned to incoming signals. Conversely, if a neuron has been increasingly active previously, it is able to decrease its output in response to later input. How does a neuron do this?

In the introduction, we have seen that the excitability of a neuron depends on, among other factors, how many voltage-gated ion channels are expressed in the membrane. Mainly the voltage-gated Na^+, K^+, and Cl^- ion channels play a role in generating action potentials. To make a neuron depolarize (and thus make it more likely to fire, or enhance excitability), the Na^+ channels are important, whereas the K^+ and Cl^- channels tend to hyperpolarize the membrane potential of a neuron (and thus make it less likely to fire). You could simplify this by saying that in most cases, Na^+ channels are excitatory K^+, and Cl^- channels are inhibitory.

Interestingly, it was shown that neurons can adapt the amount of voltage-gated ion channels in the membrane, depending on external input. This was demonstrated with electrophysiological recordings of isolated neurons in vitro: single neurons were isolated in a petri dish, and their activity was recorded with an electrode. Without a surrounding neuronal network, the neurons did not receive any synaptic input. The experimenters call this "chronic inhibition" (**Figure 2**) to indicate that neuronal activity was inhibited for a longer period of time. When the researchers probed the neuron (gave a stimulus), the isolated, chronically inhibited neurons were silent at first. But after a couple of days, the neurons became more active: the experimenters found that there were more excitatory ion channels expressed in the membrane after a couple of

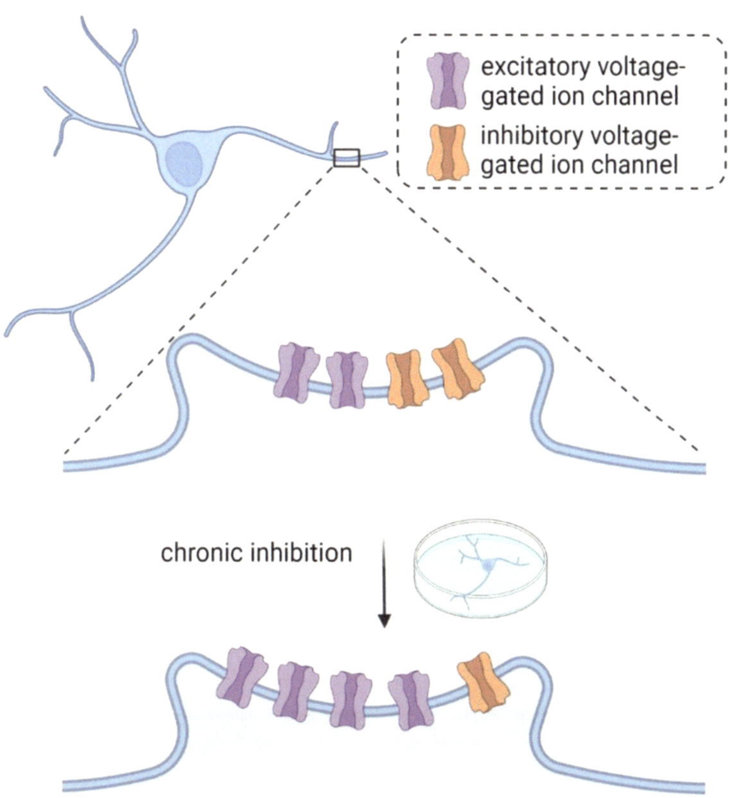

Figure 2.
(created with BioRender.com): the effect of chronic inhibition in isolated neurons in vitro. Under normal conditions, there is a balanced expression of inhibitory (orange) and excitatory (purple) voltage-gated ion channels. Following prolonged inhibition, the expression of excitatory ion channels is enhanced, and that of inhibitory ion channels reduced. Based on experiments described in [2, 3].

days in vitro, and less inhibitory channels. Also, the responsiveness of these ion channels was altered. Together this led to enhanced activity (more action potentials) of the neuron. So although initially the neurons were silent, their excitability increased again, until the level of excitability that is typically seen in that type of neuron under normal conditions (within a neural network) [3]. Importantly, when the experiments were repeated but the activity of Ca^{2+} was blocked, neurons were no longer able to do this. Apparently, individual neurons restore their excitability by using a calcium-dependent mechanism to adjust their voltage-gated ion channel expression (see **Figure 2** for a simplified schematic illustration of this process).

We have now seen an example of how neurons can restore their baseline activity levels by altering their voltage-gated ion channels. In addition to inserting more of a specific ion channel into their membrane, as we have just discussed, a neuron has other options, for example: making an ion channel more or less prone to open at certain membrane voltages by changing the composition of an ion channel. It was also shown by several studies that neurons compensate for the loss of function of one type of ion channel by inserting another type of channel with a similar function, to restore their intrinsic firing properties. These are all forms of "rebalancing ion channel distribution" (reviewed in [2]) that aim to restore the characteristic intrinsic firing properties of a neuron. The fact that a neuron is doing this, even when it is isolated in a petri dish, indicates that these firing properties must be somehow encoded genetically. This enables, but also drives a neuron to always restore its excitability to a "baseline" level.

Further, studies have shown that a neuron can move or adjust the size of its *axon initial segment*, which is the area at the beginning of an axon where a neuron generates its action potential. How far the axon initial segment is from the cell body (soma) is an important determinator for the excitability of a neuron (see [4]). Research has shown that in response to excessive input, neurons can move the axon initial segment away from the soma [5]. This too is a calcium-dependent process, and the consequence is an increased threshold to generate an action potential. Note that the reverse is also thought to be possible [6].

We have now seen a couple of examples of mechanisms in homeostatic intrinsic plasticity. It is not necessary for a neuron to receive synaptic input to trigger these mechanisms. Rather, they are driven by the membrane properties of the neuron itself. That is why it is called homeostatic intrinsic plasticity. Calcium plays an important role in homeostatic intrinsic plasticity, but the exact mechanisms through which this all happens remain to be unraveled. For homeostatic synaptic plasticity, however, the underlying processes have been identified in a bit more detail, as we will see in the next section.

2.2 Homeostatic synaptic plasticity

In the previous section, we have seen that neurons use homeostatic *intrinsic* plasticity to regulate and restore their own, intrinsic activity levels, mostly by tuning their *voltage-gated* ion channels. Next, we will explore homeostatic *synaptic* plasticity mechanisms that, as the term suggests, involve synaptic effectors: *ligand-gated* ion channels, neurotransmitters, and their receptors.

2.2.1 Postsynaptic regulation of receptors and synaptic scaling

In addition to their voltage-gated ion channels, neurons are also known to tune their ligand-gated ion channels and neurotransmitter receptors in response to network activity. You may remember reading something similar in the introduction of this chapter, where the cellular mechanisms behind Hebbian learning were briefly discussed. Like Hebbian plasticity, homeostatic synaptic plasticity involves inserting or removing receptors to and from the membrane. But in contrast to Hebbian plasticity, the effect is not reinforcing the input a neuron receives, but rather, counterbalancing it. It might seem that Hebbian and homeostatic synaptic plasticity are opposing each other—but luckily, homeostatic mechanisms do not interfere with learning, as we will see next.

As you may recall, LTP and LTD are synapse-specific processes. In contrast, homeostatic synaptic plasticity can up- or downregulate the excitability of the whole neuron at once. This explains why homeostasis does not prevent learning: if the overall excitability of the neuron is in- or decreased, the relative strength of individual synapses is unaffected. The first type of homeostatic synaptic plasticity that we will discuss does exactly that: the neuron's *total* responsiveness is scaled up or down, while each synapse maintains its relative "weight." Therefore, this type of homeostatic plasticity was termed *synaptic scaling* [7].

The first demonstrations of synaptic scaling come from experiments in vitro, where one neuron was recorded and the surrounding network activity was manipulated [7, 8]. The researchers found that blocking inhibitory signaling, which enhances the network activity, made the recorded neuron less responsive. Also, a decrease in AMPA receptors was observed. Conversely, when excitatory network activity was blocked, the neuron showed increased responsiveness and an accumulation of AMPA receptors. Moreover, a change in the composition of the AMPA receptors was demonstrated: more of the subunit GluA1 was inserted, which increased the neuron's excitability [8].

In 2013, synaptic scaling was demonstrated *in vivo*: experimenters visually deprived rats on one side (blindfolded one eye) and recorded from neurons that usually received visual input from that eye. These neurons would, after an initial silent period, resume their firing, and re-established the baseline firing rate that was typical for those neurons [9]. This experiment shows that there is compensatory tuning of neuronal activity, dependent on experience.

Since 1998, studies have demonstrated synaptic scaling in response to altered network activity, mostly by measuring changes in AMPA receptors, or the currents that are associated with these receptors: AMPA-R-mediated currents (reviewed in [2, 6]). That is not to say that the AMPA-R always changes in the same way during synaptic scaling. Rather, different circumstances evoke changes in specific receptor subunits, and each subunit has a different effect on the receptor function. For instance, the insertion of the GluA1 subunit makes the AMPA receptor more permeable to Ca^{2+}, which then increases the excitability of the synapse, but other subunits have different effects.

Synaptic scaling is a delicate and fine-tuned system with a high variety of effectors and mechanisms. And to make matters more complicated: while it was first described as a neuron-wide effect, it actually *can* occur at separate synapses, as we will see next.

Synaptic homeostatic plasticity has been shown to occur selectively at single synapses, given very localized or distinct synaptic input. As an example, consider this study in optic neurons in tadpoles. The optic neurons receive visual input from optic nerves, or pressure (vibration) input from mechanosensory nerves, at different synapses. Using localized electrophysiological recordings, the experimenters could measure the amplitude of the evoked currents in response to input from either synapse. They found that depending on the type of input, different synapses displayed homeostatic changes in current amplitudes [10]. In an experiment with cultured neurons, it was even demonstrated that two adjacent dendritic spines can independently apply postsynaptic homeostatic plasticity: one of the spines received input from a neuron with inhibited activity (due to overexpression of a K^+ channel), and only that spine showed an increase in AMPA-R-mediated current in response—to compensate for this reduced activity [11].

In the previous paragraph we have seen examples of very localized synaptic scaling. Although in some of these processes gene transcription is involved, it is not required for all types of localized homeostatic synaptic plasticity. Within dendritic spines, local protein translation allows for the rapid up-or down-regulation of receptors without the involvement of gene transcription. It is therefore not needed for the neuron to arrange this centrally in the nucleus. However, when synaptic scaling occurs in the whole cell, it usually involves gene transcription, increasing the timescale to days [12].

When it comes to the mechanisms involved in postsynaptic homeostatic plasticity, many intracellular signaling pathways are involved. Similar to LTP and LTD, intracellular calcium signaling plays an important role. In response to synaptic input, Ca^{2+} enters a neuron through a Ca^{2+}-permeable ion channel (either voltage- or ligand-gated) or is released from intracellular stores upon activation of a second messenger. When the intracellular Ca^{2+} levels rise, it serves as a starting point for molecular signaling pathways. The extent of the Ca^{2+} rise is related to the strength of the synaptic input, so Ca^{2+} levels serve as an indicator for the level of activity. How does Ca^{2+} then activate homeostatic processes?

Many different routes and effects have been demonstrated in experiments, that all rely on different signaling pathways and protein kinases, which we have introduced briefly in the introduction. To illustrate (an important) one, consider the example where there is a drop in network activity. This leads to decreased Ca^{2+} signaling in the soma of the postsynaptic neuron. As a result, there is reduced activation of a specific Ca^{2+}-dependent kinase: CaMKIV. Under normal conditions, CaMKIV regulates the transcription of certain AMPA-R subunits. Because of the drop in Ca^{2+} and the subsequently reduced functioning of CaMKIV, the transcription that is normally regulated is now enhanced, which leads to increased production of specific AMPA-R-subunits (in this case: primarily GluA2). Also, it leads to the production of more vesicles that transport these subunits to the postsynaptic membrane [13]. For a simplified visualization of this process, see **Figure 3**. As you might have guessed, this leads to enhanced AMPA-R-mediated currents, and thus, increased excitability. This is just one example of a Ca^{2+}-mediated mechanism in homeostatic synaptic plasticity in the postsynapse.

Depending on which kinases or signaling pathways are involved, a change in Ca^{2+} levels may trigger different kinds of effects. We have just seen how it can affect the transcription of genes that affect receptor expression and—functioning, but also influences the trafficking (transport) of the receptor (subunits) toward the membrane. But Ca^{2+} signaling can also have affects without involving gene transcription. As we have just discussed, single dendritic spines can also undergo synaptic scaling and for this, only local protein translation is required. In addition to the insertion of AMPA receptor subunits in the postsynapse, this may also have effects on other proteins that indirectly influence receptor function. An example of such local effects involves specialized proteins that are present in the postsynaptic membrane: transmembrane AMPA-receptor Regulating Proteins (TARPs). They are operated by a different calmodulin protein kinase, CaMKII. CaMKII can phosphorylate (and hereby

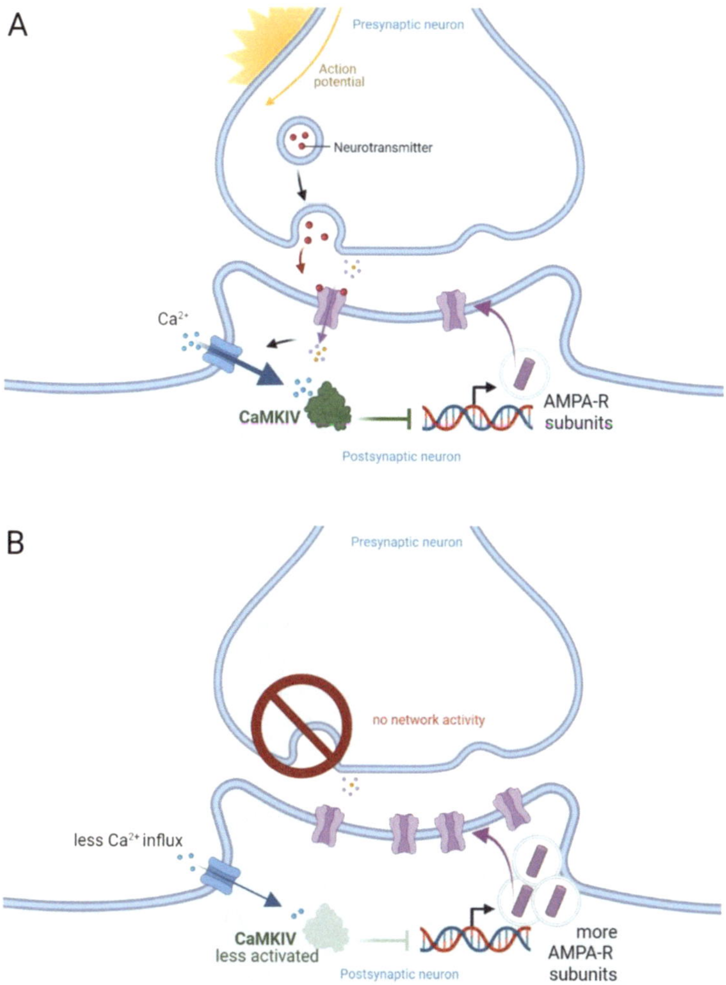

Figure 3.
(created with BioRender.com): an example of calcium-mediated homeostatic synaptic plasticity in the postsynapse. A) synaptic transmission triggers Ca^{2+} influx in the postsynapse, which activates Ca^{2+}/calmodulin-dependent protein kinase IV (CaMKIV). This kinase normally regulates the transcription of AMPA-receptor subunits. B) when there is a lack of network activity, and thus, less Ca^{2+} influx, CaMKIV is less activated. This causes enhanced transcription of AMPA-receptor subunits and their trafficking to the membrane, leading to more responsive AMPA receptors in the postsynaptic membrane.

activate) TARPS, which (as the name suggests) regulate AMPA-R, by connecting AMAPA-R to scaffold proteins such as PSD (postsynaptic density)-95. PSD acts as a sort of "anchor" for the AMPA-R; it keeps the receptor inserted. This is an example of stabilizing a receptor, which can increase the functioning of the receptor and thereby, enhance the responsiveness of the neuron. In addition to PSD-95, several other postsynaptic density proteins have been identified to play a role in homeostatic signaling (as reviewed in [14]).

In sum, neurons can counterbalance their excitability by strengthening or weakening synapses postsynaptically, either locally or globally, and this affects mostly the function or expression of (AMPA) receptors. But as we will see in the next section, homeostatic plasticity is not restricted to the postsynapse.

2.2.2 Presynaptic regulation of neurotransmitter release

Synaptic signaling does not only rely on receptor functioning in the postsynaptic neuron. The amount of neurotransmitter released by the presynaptic neuron is of course also a crucial determinant of neuronal activity. As we will see in this section, homeostatic synaptic plasticity can also occur presynaptically, by influencing the release of neurotransmitter (exocytosis).

Most of the work on presynaptic homeostatic synaptic plasticity was done in Drosophila (fruit flies), in the neuromuscular junction, which is where a neuron innervates a muscle cell. Here, it was first shown that when there was less activity in a muscle cell, the neuron that innervated this muscle cell excreted more neurotransmitters [15]. Apparently, a homeostatic mechanism takes place at the presynapse that compensates for reduced activity at the postsynapse. This implies that the presynaptic neuron has to receive a feedback signal that "informs" the presynaptic cell on the level of activity at the synapse. How does this work? First, it has been shown that presynaptic autoreceptors play an important role herein. As the term suggests, presynaptic autoreceptors are receptors that are located on the presynaptic neuron and are activated by the neurotransmitter that is produced by the very same neuron (see **Figure 1**). Thus, the amount of activated presynaptic autoreceptors is indicative of the activity levels in the synapse. Consequently, the neuron can regulate neurotransmitter release of their own neuro to adjust or restore synaptic activity. Second, it is thought that the feedback signal occurs via retrograde messengers. Retrograde messengers are compounds that are released by dendrites (so, postsynaptically) in response to strong input. They travel back (retrograde) through the synapse, where they bind to receptors that, when activated, influence neurotransmitter release (examples of these retrograde messengers and related mechanisms are reviewed in [16]).

Just like postsynaptic mechanisms, homeostatic presynaptic plasticity is also largely dependent on Ca^{2+}. Retrograde messengers and the activation of presynaptic autoreceptors result in changes in Ca^{2+} influx into the presynaptic neuron. As you may remember from the introduction, intracellular Ca^{2+} signaling regulates exocytosis: the fusion of vesicles containing neurotransmitters with the membrane, upon which the neurotransmitter is released into the synapse. Changes in the Ca^{2+} levels are therefore a direct trigger for increased or decreased neurotransmitter release into the synapse, and this does not necessarily involve gene transcription (see [17]). To summarize, the presynaptic regulation of neurotransmitter release provides a rapid mechanism that contributes to homeostatic synaptic plasticity.

2.3 Network-level homeostasis: the role of glia

So far we have discussed homeostatic mechanisms in pre- and postsynaptic neurons. But as you know, neurons are not the only type of cells that make up a neuronal network. The other type is the *glial cells*, or glia. Glia come in many different subtypes, and serve many functions: they play a role in the brain's immune response (neuroinflammation), remove cellular waste products, and guide neuronal growth and development. Glia have long been called the "supporting cells," but that would be an understatement; communication between neurons in the brain could not take place without glia.

Astrocytes are a specific subtype of glia. In addition to their role in neuroinflammation, they maintain the integrity of the blood–brain barrier. In neuronal networks, they form connections with other astrocytes and with neurons. Many synapses are surrounded by astrocyte processes, which is why the term "tripartite synapse" is sometimes used—this indicates the presence of a presynaptic neuron, a postsynaptic neuron, and an astrocyte. As you may recall from the introduction, astrocytes are known to take up excess ions (primarily K^+) after neuronal signaling, removing them from the synaptic cleft (see **Figure 1**). Through this, they have a "buffering" function and help to keep ion distribution in balance. They do the same for neurotransmitters, for example, astrocytes take up excess glutamate and thereby contribute to glutamate homeostasis. These are illustrations of direct glia-mediated homeostatic mechanisms to control neuronal activity. As we will see next, there are also indirect mechanisms through which glia contribute to homeostasis.

During neurodevelopment, glia are responsible for "pruning": removal of synapses that are redundant. Through pruning, glia help to establish a stable neuronal network, where there is a balance between excitation and inhibition. When redundant synapses are not pruned, it can lead to excessive connectivity which disrupts this balance. As such, glia are critical in creating a healthy neuronal network where activity remains within functional limits.

As stated before, glia are also the mediators of neuroinflammation. Glia produce cytokines, which are signaling compounds that play a role in inflammatory reactions. Interestingly, many cytokines are also involved in neuronal signaling. For example, consider tumor necrosis factor—alpha (TNF-α), a compound with many functions in the central nervous system (see for a review). TNF-α is produced by neurons, but also (and primarily) by glia. An increase in TNF-α can trigger enhanced glutamate release and also AMPA receptor expression [18]. This too is an example of the indirect effects of glia on neuronal activity.

The examples mentioned here suggest that there is a link between neuroinflammation and the excitability of neuronal networks. In inflammatory conditions, for example caused by an infection or damage to the neural tissue, glia become reactive. This serves to battle the infection or damage, but it has a downside: reactive glia show compromised homeostatic functioning, which can have consequences for brain function, as we will discover in the next section.

2.4 Disrupted homeostasis in brain disease

If homeostasis is disrupted, neurons cannot keep their excitability within limits. As you can imagine, this will have detrimental effects on brain function. In this section, we will briefly highlight two examples of brain diseases where disturbed homeostasis is known to be involved. Bear in mind that disrupted homeostasis may actually play an important role in numerous other brain diseases—what we describe here is just the tip of the iceberg.

The first brain disease with a clear link to disrupted homeostasis is epilepsy. Epilepsy is a neurological disorder affecting millions of patients worldwide. As you probably know, the main symptom of epilepsy is epileptic seizures. During a seizure, there is abnormal, excessive, and uncontrolled firing of neurons. In epileptic brains, neuronal networks tend to be hyperexcitable—the homeostatic control of neuronal activity is apparently not functioning as it should. The challenge in epilepsy research is to find which processes cause this malfunction of homeostasis. Increased neuroinflammation is suspected to play an important role here (although many other factors are thought to contribute as well—see for a review [19]). To illustrate the link between neuroinflammation and increased excitability, consider the role of astrocytes. In response to damage or infection (or other inflammatory circumstances), astrocytes get activated: they become "reactive" to respond to the damage or infection, as was already mentioned in the previous section. When astrocytes become reactive, their glutamate and K^+ buffering capacities decrease. To illustrate this, an experiment showed that there is decreased expression of the Excitatory Amino Acid Transporters 1 and 2 (EAAT-1 and EAAT-2) in epilepsy patients. EAAT-1 and EAAT-2 are glutamate transporters that are normally found on glia—they mediate the glutamate buffering function of astrocytes. In this research, it was also shown that decreased EAAT expression coincides with astrocytes being reactive [20]. When glutamate is not sufficiently taken up by astrocytes, it is available in synapses much longer—which can lead to more AMPA-R-mediated signaling and thus, enhanced excitatory neuronal activity. Of note, reactive astrocytes are not only seen in epilepsy, but also in other brain diseases.

Disrupted homeostasis is thought to contribute to the pathology in several psychiatric illnesses as well. For example, the malfunction of presynaptic dopaminergic autoreceptors has been shown to play a role in schizophrenia and ADHD. To briefly explain this: the autoreceptors on dopaminergic cells normally inhibit the release of dopamine when they are activated—they are key in the feedback loop that keeps dopamine signaling within limits (review section 2.2.2 for a description of presynaptic autoreceptors). In mouse models and in human patients, it has been shown that this feedback mechanism is not functioning properly, which leads to excessive dopamine release [21]. Excessive dopamine signaling is known to underlie schizophrenia, but may also play a role in ADHD.

These are only a couple of examples showing the possible consequences of disrupted homeostatic control of neuronal activity, and they clearly illustrate the importance of homeostasis in the brain. Restoring homeostasis may be (at least a big part of) the solution for many diseases but is extremely difficult to achieve.

There is no treatment that selectively tunes one process to reinforce homeostasis in a specific neuronal network, nor is there a "reset-button" to restore homeostasis once it has been disrupted. Throughout this chapter, we have briefly and globally addressed some of the mechanisms that control neuronal activity. These (and other, non-discussed) mechanisms all work simultaneously to produce and maintain homeostasis, and often, interfering with one also has consequences for another. As you may have gathered from this chapter, homeostasis is a very intricate and sensitive set of mechanisms that is crucial for healthy brain functioning.

3. Discussion

The human brain contains billions of neurons. Every neuron communicates with thousands of other neurons, so there are trillions of connections within the brain. The connections within and between neuronal networks are formed and shaped by neuronal activity, which means that neuronal activity is subject to, but also causes, a high degree of plasticity. This plasticity is critical for learning, but it also poses a challenge: neuronal activity needs to be limited to remain functional. Excessive neuronal firing hampers information transfer and could lead to epileptic seizures. On the other hand, if neurons are not active at all, you would not be able to think, feel, or act. Homeostatic mechanisms ensure that neuronal activity remains within a functional range. In this chapter, we have discussed the basics of *homeostatic control of neuronal activity.*

Homeostatic control of neuronal activity aims to restore neuronal excitability to baseline. Interestingly, homeostatic mechanisms also depend on plasticity in the brain. We distinguish two types of homeostatic plasticity: homeostatic intrinsic plasticity and homeostatic synaptic plasticity. In homeostatic intrinsic plasticity, neurons tune their own intrinsic firing properties, for example by adjusting voltage-gated ion channels, as we have seen in section 2.1.

Homeostatic synaptic plasticity targets neurotransmitter-dependent signaling. In section 2.2.1, we discussed synaptic scaling: a neuron strengthens or weakens its connections by adapting its receptor expression or adapting the receptor subunit composition. In addition to postsynaptic mechanisms, homeostatic synaptic plasticity can also be presynaptic, such as the control of neurotransmitter release by the activation of presynaptic autoreceptors, discussed in section 2.2.2.

In section 2.3, we have discussed the role of the glia in the homeostatic control of neuronal activity, from which we can understand the link between neuroinflammation and disrupted homeostasis. Disrupted homeostasis contributes to brain diseases such as epilepsy, or psychiatric illnesses, as we have seen in section 2.4.

An interesting proposition is that these (and other) brain disorders can arise because of the high degree of plasticity in neuronal networks. In other words, maybe there is a price we pay for our adaptive brains.

If brain disorders are the other side of the coin of neuronal plasticity, where does that leave homeostatic plasticity? As we have seen throughout this chapter, the synaptic plasticity that enables learning (LTP and LTD; we call this Hebbian plasticity) shares molecular mechanisms with homeostatic plasticity. It has been proposed that Hebbian and homeostatic plasticity are basically the same process, but with different outcomes: for learning, the plasticity is used to reinforce activity, whereas for homeostasis it is used to compensate activity. If they use the same (or highly similar) molecular mechanisms, it is still unclear how and why neurons can display different forms of plasticity, with different outcomes. Another question then is: why does homeostasis not counteract learning? Future research will undoubtedly address these complicated questions.

In this chapter, we have attempted to give you a basic overview of homeostatic control of neuronal activity. We hope it has inspired you to explore this topic further; there is much more to learn about homeostasis in the brain.

Author details

Cato Drion
University of Groningen, Groningen, The Netherlands

*Address all correspondence to: c.m.drion@rug.nl

References

[1] Malenka RC, Bear MF. Review LTP and LTD: An embarrassment of riches. Neuron. 2004;**44**:5-21

[2] Davis GW. Homeostatic signaling and the stabilization of neural function. Neuron. 2013;**80**(3):718-728 Available from: https://linkinghub.elsevier.com/retrieve/pii/S089662731300901X

[3] Turrigiano G, Abbott LF, Marder E. Activity-dependent changes in the intrinsic properties of cultured neurons. Science. 1994;**264**(5161):974-977. DOI: 10.1126/science.8178157

[4] Leterrier C. The axon initial segment: An updated viewpoint. The Journal of Neuroscience. 2018;**38**(9):2135-2145. Available from: DOI: 10.1523/JNEUROSCI.1922-17.2018

[5] Grubb MS, Burrone J. Activity-dependent relocation of the axon initial segment fine-tunes neuronal excitability. Nature. 2010;**465**(7301):1070-1074 Available from: http://www.nature.com/articles/nature09160

[6] Turrigiano G. Too many cooks? Intrinsic and synaptic homeostatic mechanisms in cortical circuit refinement. Annual Review of Neuroscience. 2011;**34**(1):89-103. DOI: 10.1146/annurev-neuro-060909-153238

[7] Turrigiano GG, Leslie KR, Desai NS, Rutherford LC, Nelson SB. Activity-dependent scaling of quantal amplitude in neocortical neurons. Nature. 1998;**391**(6670):892-896 Available from: http://www.nature.com/articles/36103

[8] O'Brien RJ, Kamboj S, Ehlers MD, Rosen KR, Fischbach GD, Huganir RL. Activity-dependent modulation of synaptic AMPA receptor accumulation. Neuron. 1998;**21**(5):1067-1078. Available from: https://linkinghub.elsevier.com/retrieve/pii/S0896627300806248

[9] Hengen KB, Lambo ME, Van Hooser SD, Katz DB, Turrigiano GG. Firing Rate Homeostasis in visual cortex of freely behaving rodents. Neuron. 2013;**80**(2):335-342. Available from: https://linkinghub.elsevier.com/retrieve/pii/S0896627313007988

[10] Deeg KE, Aizenman CD. Sensory modality–Specific homeostatic plasticity in the developing optic tectum. Nature Neuroscience. 2011;**14**(5):548-550 Available from: http://www.nature.com/articles/nn.2772

[11] Béïque JC, Na Y, Kuhl D, Worley PF, Huganir RL. Arc-dependent synapse-specific homeostatic plasticity. Proceedings of the National Academy of Sciences. 2011;**108**(2):816-821. DOI: 10.1073/pnas.1017914108

[12] Sutton MA, Taylor AM, Ito HT, Pham A, Schuman EM. Postsynaptic decoding of neural activity: eEF2 as a biochemical sensor coupling miniature synaptic transmission to local protein synthesis. Neuron. 2007;**55**(4):648-661. Available from: https://linkinghub.elsevier.com/retrieve/pii/S0896627307005752

[13] Ibata K, Sun Q, Turrigiano GG. Rapid synaptic scaling induced by changes in postsynaptic firing. Neuron. 2008 [cited 2022 Sep 5];**57**(6):819-826. Available from: https://linkinghub.elsevier.com/retrieve/pii/S0896627308002134

[14] Fernandes D, Carvalho AL. Mechanisms of homeostatic plasticity in the excitatory synapse. Journal of Neurochemistry. 2016;**139**(6):973-996. Available from:. DOI: 10.1111/jnc.13687

[15] Davis GW, Goodman CS. Genetic analysis of synaptic development and plasticity: Homeostatic regulation of synaptic efficacy. Current Opinion in Neurobiology. 1 Feb 1998;**8**(1):149-156. Available from: https://linkinghub.elsevier.com/retrieve/pii/S0959438898800184

[16] Citri A, Malenka RC. Synaptic plasticity: Multiple forms, functions, and mechanisms. Neuropsychopharmacology. 2008;**33**(1):18-41 Available from: http://www.nature.com/articles/1301559

[17] Frank CA, Kennedy MJ, Goold CP, Marek KW, Davis GW. Mechanisms underlying the rapid induction and sustained expression of synaptic homeostasis. Neuron. 2006;**52**(4):663-677. Available from: https://linkinghub.elsevier.com/retrieve/pii/S0896627306007367

[18] Heir R, Stellwagen D. TNF-mediated homeostatic synaptic plasticity: From in vitro to in vivo models. Frontiers in Cellular Neuroscience. 2020 Sep;**30**:14 Available from: https://www.frontiersin.org/article/10.3389/fncel.2020.565841/full

[19] Boison D, Sandau US, Ruskin DN, Kawamura M, Masino SA. Homeostatic control of brain function – New approaches to understand epileptogenesis. Frontiers in Cellular Neuroscience. 2013;**7**:1-12. Available from: http://journal.frontiersin.org/article/10.3389/fncel.2013.00109/abstract

[20] Sarac S, Afzal S, Broholm H, Madsen FF, Plough T, Laursen H. Excitatory amino acid transporters EAAT-1 and EAAT-2 in temporal lobe and hippocampus in intractable temporal lobe epilepsy. APMIS. 2009;**117**(4):291-301. DOI: 10.1111/j.1600-0463.2009.02443.x

[21] Gowrishankar R, Gresch PJ, Davis GL, Katamish RM, Riele JR, Stewart AM, et al. Region-specific regulation of presynaptic dopamine homeostasis by D_2 autoreceptors shapes the *in vivo* impact of the neuropsychiatric disease-associated DAT variant Val559. The Journal of Neuroscience. 2018;**38**(23):5302-5312. DOI: 10.1523/JNEUROSCI.0055-18.2018

Chapter 4

Cholesterol Homeostasis, Mechanisms of Molecular Pathways, and Cardiac Health

Rajiv Kumar, Neelam Chhillar, Dhruv Sanjay Gupta, Ginpreet Kaur, Shailey Singhal, Tanya Chauhan and Sandeep Mittan

Abstract

The metabolism of lipoproteins, which regulate the transit of the lipid to and from tissues, is crucial to maintaining cholesterol homeostasis. Cardiac remodeling is referred to as a set of molecular, cellular, and interstitial changes that, following injury, affect the size, shape, function, mass, and geometry of the heart. Acetyl coenzyme A (acetyl CoA), which can be made from glucose, amino acids, or fatty acids, is the precursor for the synthesis of cholesterol. In this article, authors explain concepts behind cardiac remodeling, its clinical ramifications, and the pathophysiological roles played by numerous various components, such as cell death, neurohormonal activation, oxidative stress, contractile proteins, energy metabolism, collagen, calcium transport, inflammation, and geometry. The levels of cholesterol are traditionally regulated by two biological mechanisms at the transcriptional stage. First, the SREBP transcription factor family regulates the transcription of crucial rate-limiting cholesterogenic and lipogenic proteins, which in turn limits cholesterol production. Immune cells become activated, differentiated, and divided, during an immune response with the objective of eradicating the danger signal. In addition to creating ATP, which is used as energy, this process relies on metabolic reprogramming of both catabolic and anabolic pathways to create metabolites that play a crucial role in regulating the response. Because of changes in signal transduction, malfunction of the sarcoplasmic reticulum and sarcolemma, impairment of calcium handling, increases in cardiac fibrosis, and progressive loss of cardiomyocytes, oxidative stress appears to be the primary mechanism that causes the transition from cardiac hypertrophy to heart failure. De novo cholesterol production, intestinal cholesterol absorption, and biliary cholesterol output are consequently crucial processes in cholesterol homeostasis. In the article's final section, the pharmacological management of cardiac remodeling is explored. The route of treatment is explained into different steps: including, promising, and potential strategies. This chapter offers a brief overview of the history of the study of cholesterol absorption as well as the different potential therapeutic targets.

Keywords: cholesterol homeostasis, cardiac remodeling, inflammation, atherosclerotic, cardiomyocytes, atherosclerosis, catabolic and anabolic pathways

1. Introduction

> *A self-regulating phenomenon called homeostasis facilitates biological systems to sustain stability while answering the environmental influences best for its survival. If homeostasis is succeeded, life carries on; if it is not, heartbreak or death comes to an end.*

Dysregulated cholesterol has been emphasized as a primary factor in atherosclerosis, which is the leading cause of mortality in developed countries globally. Although the cell needs cholesterol, its presence exceeding the required amount can be damaging, therefore preserving cholesterol homeostasis is vital for achieving a healthy cellular function and outline. Atherosclerotic heart disease risk has long been connected with elevated levels of plasma cholesterol [1]. Additionally, cholesterol participates in various cellular processes, including synthesis of hormones, and vitamin D, maintenances cell membranes and cell networks of brain. However, as blood cholesterol levels increase, they can occasionally show damaging effects on physiology [2]. Cholesterol homeostasis is directly related to lipoprotein metabolism, which facilitates lipid transport to and from the organs. A summary of several physiological routes and paths has been discussed that controls the intestinal absorption of cholesterol, its generation, as well as uptake. Regulation of transintestinal transport and transfer of cholesterol, which is decisive for preserving cholesterol homeostasis and avoiding atherosclerosis, is an eye-catching and important research theme today [3]. By offering an overview of the key routes and proteins that govern lipoprotein homeostasis, this study adds to considerable attention that has recently been explored and reported in this area [4]. Further, this chapter sheds light on the interface between central and peripheral metabolism of cholesterol and its role in health and disease [5]. The authors have discussed current findings on cholesterol physiology, with a focus on cholesterol production, fecal excretion, cholesterol absorption, and new (potential) therapy approaches for the management of hypercholesterolemia [6]. The ubiquitous, multistep physiological transformation is defined as the cell cycle, which is crucial for the propagation and division of cells. Although several cytoplasmic and nuclear regulators have been reported and revealed, the physiological importance of membrane lipids in cell cycle regulation has not been exhaustively examined.

In the past few decades, scientists have tried exploring signaling pathways that control cholesterol homeostasis, the correlation between plasma cholesterol levels and interconnected processes responsible for initiating cardiovascular disease [7]. Besides this, the association between plasma cholesterol concentrations and the routes through cholesterol homeostasis occurs have been explored. The authors have specifically examined the needs of strict cholesterol regulation for controlling cell cycle progression and emphasized on the overall functioning of the membrane in regulating cholesterol physiology and the cell cycle [8]. To accomplish this, both proximal and distal inhibitors of cholesterol synthesis were studied in detail. Further, authors highlighted its impact on cell cycle progression [9]. These newly explored phenomena are pertinent in the context of illnesses such as cancer, atherosclerosis, and neurodegenerative conditions such as Alzheimer's disease, which have demonstrated a direct linkage to abnormalities in cholesterol homeostasis and biosynthesis [10]. Therefore, to understand the regulation of optimal cellular and systemic operations, a comprehensive look at cholesterol homeostasis is essential (**Figure 1**).

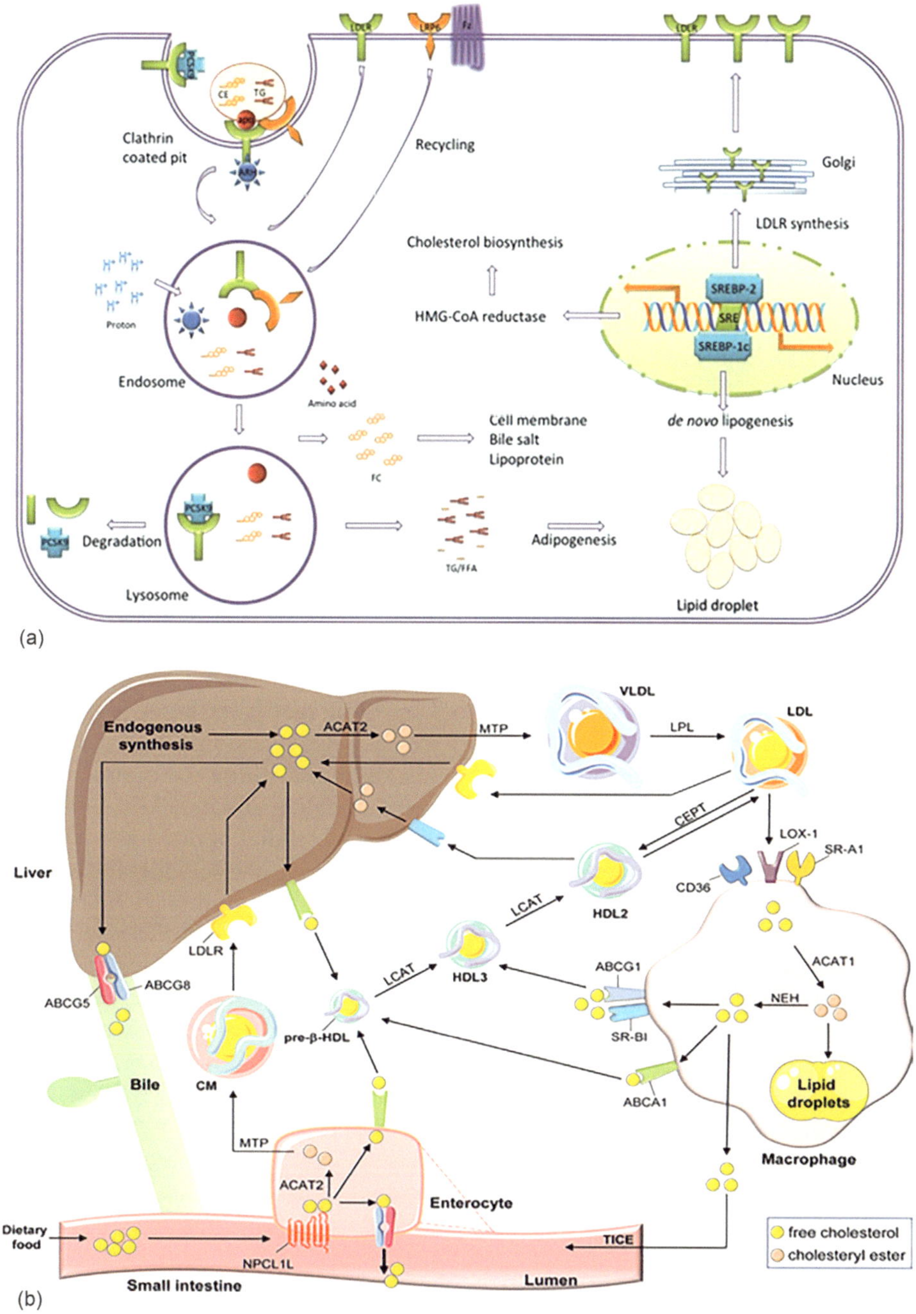

Figure 1.
(a) Cholesterol homeostasis throughout the cell. Lipoprotein and lipid metabolism depends on the uptake of lipoproteins by vesicles. Reprinted (adapted) with permission from [11]. Copyright (2012) Arya Mani et al. Yale J Biol Med., PMC. Pub. Med. Central. (b) Transport of cholesterol, resource of human cholesterol, and protein mediates dietary free cholesterol (FC) regulation. Reprinted (adapted) with permission from [12]. Copyright (2022) Jihong Han et al. Signal Transduction and Targeted Therapy, Nature.

Since cholesterol is an essential component of every cell membrane and the starting material for bile acids, lipids, steroid hormones, and lipophilic vitamins, it plays a crucial role in the functioning of human body. Additionally, it promotes cell growth, transmembrane signaling mechanisms, membrane trafficking, and maintains physiological brain functioning [13]. Cholesterol and other crucial lipids play a key role in various physiological routes and paths at cellular and systemic levels of the organism. However, too much cellular cholesterol is dangerous; therefore, sophisticated mechanisms have been established to strictly control this vital lipid [14]. This lipid is both essential and fatal, thus it's crucial to regulate its levels to sustain health. The amount of cholesterol present in the cells represents a dynamic equilibrium between its production, esterification, absorption, and export since cholesterol is transformed as neutral cholesteryl esters either by retention in vacuoles or outflow as a constituent of lipoproteins.

The most recent developments in each of the four processes of cholesterol metabolism are described in different sections of this chapter. It is argued how the functioning of these pathways is primarily ruled and how they respond to altering sterol levels. Finally, how these pathways work together to keep cholesterol levels in check has been discussed and portrayed through diagrams in the text. Moreover, the molecular mechanisms underlying cardiac remodeling and subsequent onset of disease conditions are highlighted, with a focus on the myocardial-infarction-related pathways [15]. Accordingly, the development of innovative therapeutic methodologies for treating heart failure and mitigation of cardiac consequences may be fortified by accurate knowledge about cell signaling that contributed in cardiac remodeling.

In addition to the aforementioned highlights, the authors have elaborated on different research fields to look into the domain of signaling and its role in cellular cholesterol homeostasis. Besides being a crucial part of cellular homeostasis, cholesterol is also necessary for manufacturing steroid hormones, bile acid metabolism, and cellular structures such as lipid rafts. These important tasks require that cholesterol levels in cells be retained within narrow bounds. Many common disorders, including atherosclerosis, where cellular-cholesterol accumulation prejudices regular cellular functioning, are thought to be triggered by an imbalance between uptake, synthesis, and export [16]. To identify new treatment targets, significant research is still underway, exploring the precise regulation of cholesterol homeostasis and cardiac diseases. This chapter emphasizes on the most recent advancements in studying the physiology of cholesterol homeostasis and underlying mechanisms of molecular pathways. The authors have delivered a concise summary of various topics in every section on how cholesterol influences cardiac health, inhibition of atherosclerosis and cholesterol-lowering interventions. Besides that, a separate section entitled *cholesterol homeostasis: an overview* was conferred. Molecular pathways underlying cholesterol homeostasis have been discussed separately in this chapter. Moreover, the cholesterol absorption and its metabolism, cholesterol biosynthesis, balance, efflux, enzymatic control, cell signaling, and cellular cholesterol homeostasis have been explored separately. The progression of cardiac remodeling, neuroinflammation, neurodegeneration, and cholesterol homeostasis have also been explored, to offer a deeper insight into the discussed topic [17]. Finally, cholesterol homeostasis and targeted therapeutics have been elaborated for an inference [18]. Thus, these sections of this chapter briefly cover the attempts undertaken to construct various models of the complicated mechanisms involved in cholesterol homeostasis and their applications in disease management.

2. Cholesterol homeostasis and molecular pathways: an overview

All cells in vertebrates contain cholesterol, and while the metabolism of cholesterol differs in different tissues, it remains the same in all cells overall. Animal cells obtain cholesterol both endogenously, through a well-organized process commencing with acetyl coenzyme A (acetyl-CoA), and exogenously, through the bloodstream, in the presence of apolipoprotein B-containing lipoproteins similar to minimum lipoprotein (LDL) [19]. Mechanisms governing cholesterol homeostasis are still poorly understood, despite great progress in understanding the equilibrium between cholesterol synthesis and transport. Long-standing research has linked cholesterol to numerous conditions and health issues that affect people, and although it is incontrovertibly important, very high amounts can have negative cellular effects and can even cause disorders such as atherosclerosis and type II diabetes. High hydrophobicity of cholesterol and the surrounding membrane environment, however, make it difficult to examine cholesterol homeostasis. As a result, cells have evolved sophisticated systems to control the quantity and distribution of sterols inside the cells [20]. The homeostasis of cholesterol is governed by a complicated equilibrium resulting from its synthesis, ingestion, absorption, and excretion, which is managed by lipoprotein trafficking. Intense research is being carried out on the regulation of cholesterol homeostasis, and signaling significance is progressively coming to light [21]. By using receptor-mediated endocytosis, LDL particles are transported to peripheral cells where they are digested in the lysosomes to release free cholesterol. With a focus on cholesterol absorption, cholesterol production, fecal excretion, and a new (potential) therapy approach for hypercholesterolemia, this section will address the state of current features of cholesterol physiology and provides a comprehensive overview of cholesterol homeostasis.

Feedback mechanisms that work on both the transcriptional and posttranscriptional stages precisely control the amount of intracellular cholesterol. The transcriptional activation of 3-hydroxy-3-methylglutaryl coenzyme A reductase (HMGCR), the rate-limiting enzyme of cholesterol biosynthesis, as well as nearly all downstream enzymes of the mevalonate (MVA)-pathway, is regulated by the ER-bound sterol controlling element-requisite proteins when intracellular cholesterol levels are low. The MVA-pathway governs sterol metabolism and the materialization of isopentenyl pyrophosphate and dimethylallyl pyrophosphate, which bring building blocks for the biosynthesis of chemicals applied in many cellular activities such as hormone production, protein anchoring, cell membrane maintenances, protein prenylation, and N-glycosylation [22]. Additionally, sterol regulatory element–binding proteins (SREBPs) stimulate the transcription of LDL receptors (LDLr), increasing the uptake of cholesterol by cells. Nuclear hormone receptors called liver X receptors (LXR) support the homeostasis of cholesterol. General characteristics of cholesterol metabolism and the most recent research supporting crucial function of miRNA in controlling cholesterol balance are further discussed in this chapter. There are opportunities for researchers to explore how miR-33a and miR-33b (a family of microRNA precursors, which are processed by the Dicer enzyme to give mature microRNAs) interrupt the epigenetic regulation of the physiology of cholesterol homeostasis. A vital cardiovascular disease risk factor is hypercholesterolemia. It is brought about by a deranged ratio of cholesterol absorption to its oozing into the circulation (**Figure 2**). An intricate interplay between enzymes, non-coding RNAs, transport proteins, and transcription factors controls the pathways [23]. Insight into underlying mechanisms has grown significantly over past two decades, yet there are many unanswered questions, notably concerning cholesterol transference within neighboring cells.

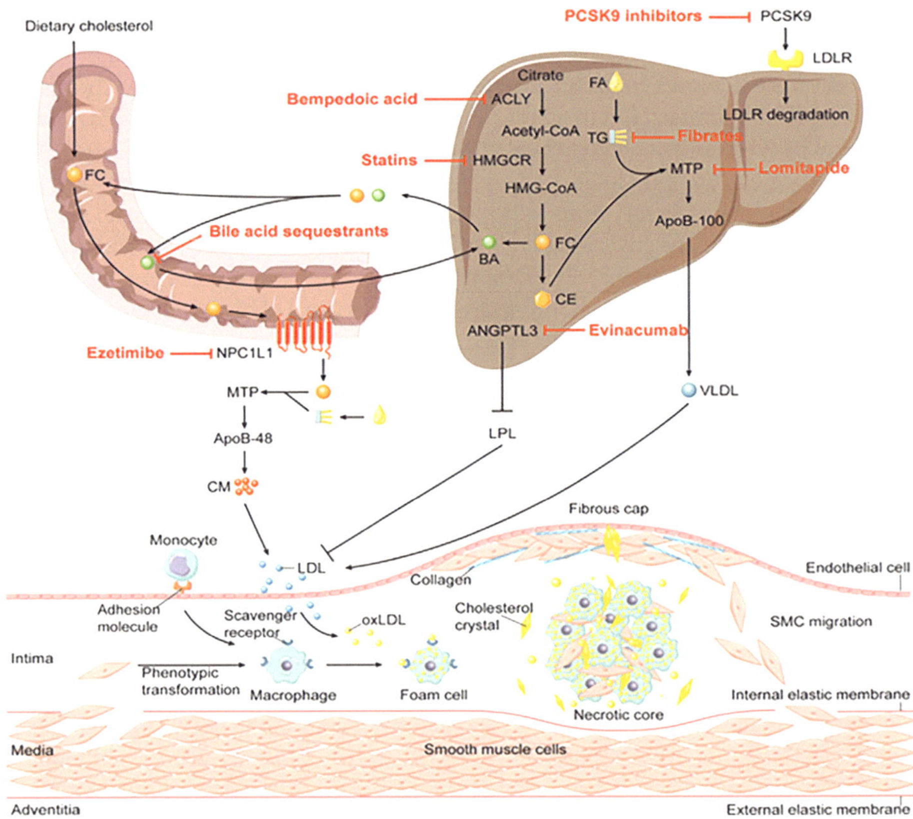

Figure 2.
Inhibition of atherosclerosis by cholesterol-lowering interventions. Cholesterol-lowering treatments that prevent atherosclerosis. By reducing the generation of acetyl-CoA and HMG-CoA, respectively, bempedoic acid and statins decrease cholesterol synthesis [12]. Reprinted (adapted) with permission from [12]. Copyright (2022) Jihong Han et al. Signal Transduction and Targeted Therapy, Nature.

Recently, the gut has drawn attention as a crucial regulating point in cholesterol homeostasis, slowly replacing the liver's endless and concerned emphasis. The major goals of this discussion are to investigate the "mysteries" of the "universe" of cholesterol homeostasis and to recommend novel, potential therapeutic targets that are emerging a methodology to offset the damaging effects of dysregulation of this route in illnesses. Additionally, mounting research has emphasized how hypercholesterolemia underwrites the development of various neurological diseases. Cholesterol-driven inflammation in patients with Alzheimer's disease and major depressive disorders seems liked to change monoamine release, membrane fluidity and permeability, vesicular trafficking, and neuroendocrine function [24]. Finally, when neurodegenerative illnesses progress, patients develop early abnormal metabolic features that precede the neurological warning signs. Even though cholesterol homeostasis serves physiological tenacity, altered cholesterol metabolism in the periphery poses a significant risk of metabolic and cardiovascular conditions, whereas, while concerning significant nerve networking, its impairment is associated with numerous neurodegenerative illnesses, as outlined above [20]. Although blood-brain barrier safeguards that cholesterol metabolism (CM) in peripheral tissues is a distinct stage that occurs in the brain, the cross talk between central and peripheral CM is at present underappreciated in both normal and pathological settings.

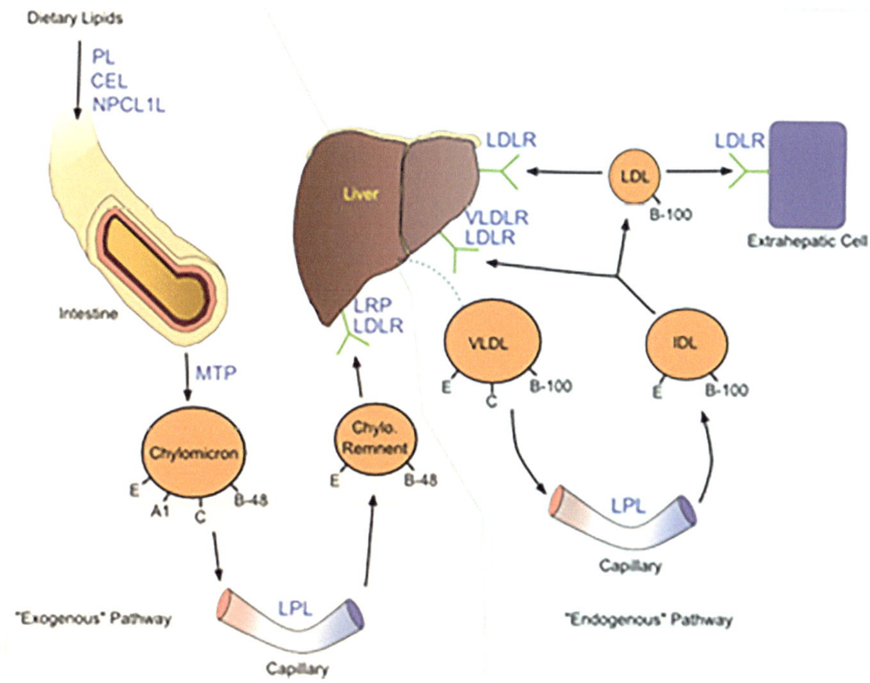

Figure 3.
A variety of lipoproteins and proteins are needed for the transport of exogenous cholesterol and de novo cholesterol [25]. Reprinted (adapted) with permission from. Copyright (2009) Zhihua Jiang et al. Int. J. Biol. Sci., Ivyspring International Publisher.

Complex uptake, efflux, and synthesis-related mechanisms govern cholesterol homeostasis. Although substantial research has been done to clarify pathways concerning cholesterol regulation (**Figure 3**), limited research articles are available that cover and explain signaling roles in these events [26]. The authors here have attempted to go over several effects that signaling has on these practices and the performance of cholesterol homeostasis progression, comprising transcriptional mechanisms.

Healthy cholesterol homeostasis is crucial for physiological functioning and can typically be considered successful when it is sustained through a dynamic equilibrium involving intake, cellular absorption, transport, production, and efflux [27]. The authors have summarized recent findings on the routes of molecular systems that govern cholesterol homeostasis and indicate areas for further study. Alongside preserving membrane fluidity and porosity, cholesterol also regulates transmembrane signaling pathways and aids in the production of vitamin D, steroid hormones, and bile acids. These cellular events are crucial for preserving membrane permeability. While many intracellular processes are influenced by cholesterol, it is generally recognized that excessive plasma cholesterol levels are the first to cause atherosclerotic heart disease [28]. Because there is strong association-involving plasma blood cholesterol and the risk of developing cardiovascular disease, clinical research has helped in clarifying the mechanisms involved in maintaining cholesterol levels and its role in the development of atherosclerosis [29]. Disrupted cholesterol homeostasis not only aids the pathogenesis of cardiovascular and cerebrovascular problems but also plays a role in the emergence of numerous other illnesses such as cancer and neurological disorders.

Atherosclerotic disease routes and metabolic syndrome are caused by low or high levels of cellular cholesterol. Mammalian cells either produce cholesterol from acetyl

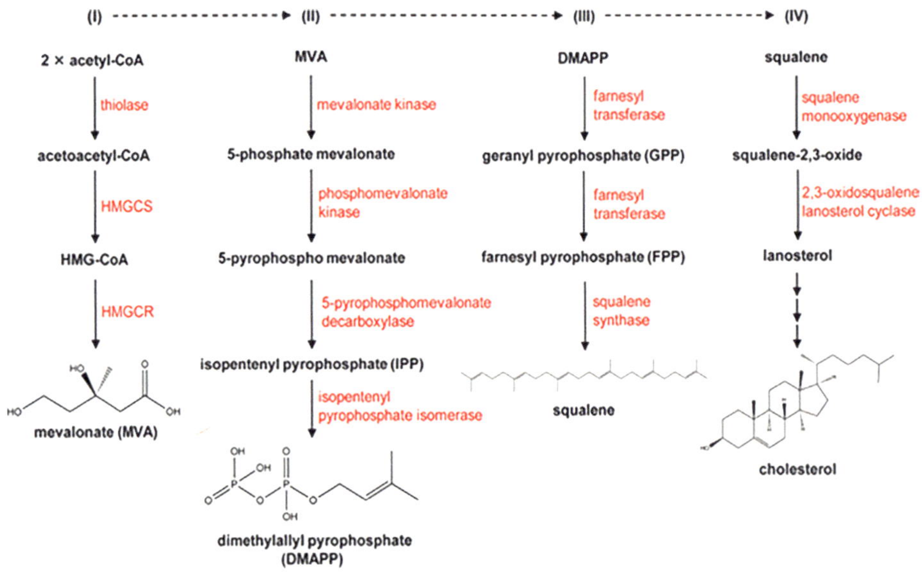

Figure 4.
The process via which cholesterol is made. All of the carbon atoms used in the production of cholesterol come from acetyl-CoA. There are four distinct phases in the production of cholesterol. Reprinted (adapted) with permission from [12]. Copyright (2022) Jihong Han et al. Signal Transduction and Targeted Therapy, Nature.

coenzyme A inside the cell or take it up from the bloodstream in the form of plasma lipoproteins (acetyl-CoA) (**Figure 4**). This route is intricately organized at different levels [19]. The various aspects and phenomena by which cellular cholesterol absorption and uptake are organized have been highlighted in this section.

3. Cholesterol absorption and metabolism

An essential biological process that maintains cholesterol homeostasis is the intestinal absorption of nutritive cholesterol. It is a necessary phenomenon, influencing the metabolic effects of cholesterol homeostasis. Bile salts play a role in the early absorption of dietary cholesterol through the creation of emulsions. The gut is highly selective to certain sterols and regulates the rate of dietary cholesterol absorption on a day-to-day basis. This distinct molecular process was recently studied and extended and approved by the researchers because of its physiological significance. Thus, every day, a large amount of cholesterol is absorbed during the digestion process [30]. Overall, roughly, several sterols in equal amounts to the 0.5-g dose of dietary cholesterol were observed to participate in this process [31]. Hepatocytes and enterocytes are the main sites of cholesterol production, even though all cells can synthesize it. Notably, despite food only giving 300–400 mg of cholesterol per day, the majority of the influence of intestinal cholesterol collection is accounted for by endogenous sources (800–1000 mg) [32].

Moreover, several plant sterols (sitosterol, avanesterol, and brassicasterol) also undergo absorption. However, it was reported that human body only allows the absorption of cholesterol and maintains its concentration in the body, and only a few plant sterols are retained during this process. This route is disrupted in a rare illness called, sitosterolemia, and is expressed in the form of genetic defects. These irregularities are responsible for the alteration in normal physiology and can be marked

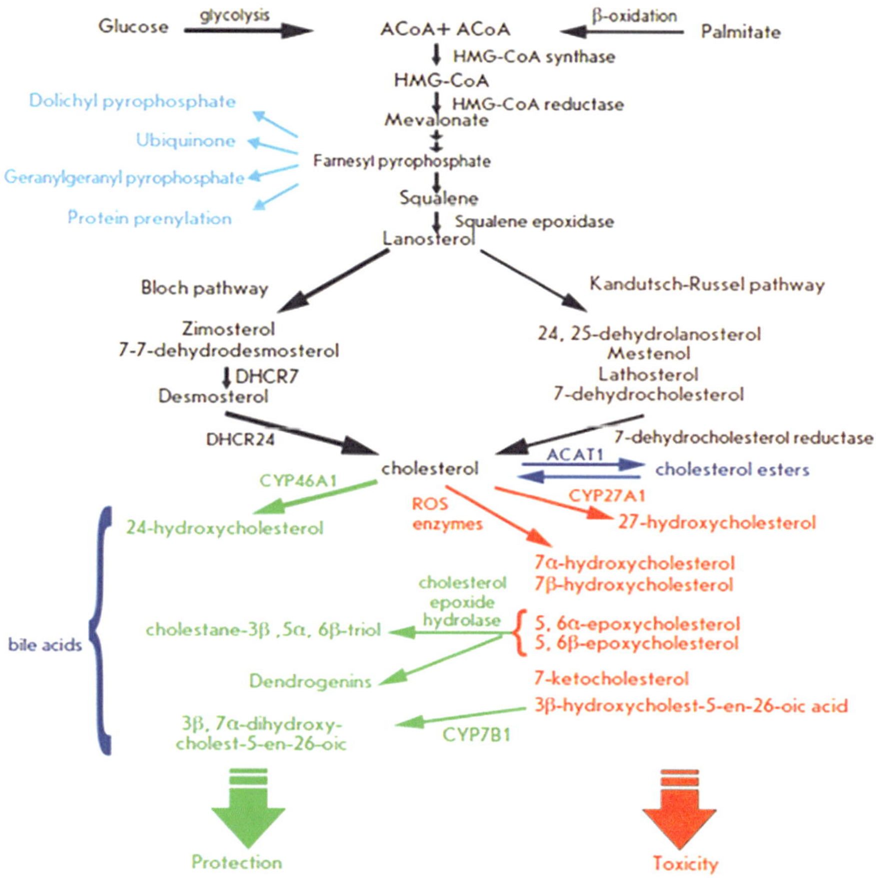

Figure 5.
Synthesis of cholesterol and production of oxysterol. Acetyl-coenzyme A is converted into cholesterol by an elaborate enzymatic mechanism. Reprinted (adapted) with permission from [34]. Copyright (2016) A. M. Petrov et al. Acta Naturae, Europe PMC.

in specific molecular pathways and identified as the main cause of these types of physiological defects. These investigations further evidenced the processes and have helped in clarifying the molecular mechanisms governing sterol absorption and excretion [33]. Several mechanisms such as micelles excreting sterols, solubility, and a monitoring protein setup participate in cholesterol absorption (**Figure 5**). These routes control sterol balance and other related activities. The protein Niemann-Pick-C1-like-1 standardizes the overall absorption of cholesterol in the small intestine [35]. The minute cholesterol molecule enters the enterocyte, post which the endoplasmic reticulum membrane enzyme ACAT2 (ER) esterifies it into fatty acid. Following its release into the bloodstream via the thoracic duct, the subsequent cholesteryl ester is combined with chylomicrons and sent to the Golgi apparatus, where it undergoes further distribution.

Additionally, a limited quantity of free cholesterol is released back into the intestinal lumen via a few apical transporters, including ATP-binding cassette transporter subfamily G member 5 (ABCG5) and member G 8 [36]. The Liver X Receptor, a nuclear receptor that inhibits NPC1L1 and activates ABCG5 and ABCG8, controls the uptake and release of intestinal cholesterol [37]. Functioning as cholesterol sensors, two Liver X Receptor isoforms-LXR (Nr1h3) and LXR (Nr1h2) have parallel effects on the production of these proteins. Additionally, LXRs promote ABCA1 expression, which further elevates cholesterol levels [38]. Actions of the Liver X Receptor together preclude cholesterol from building up in enterocytes [39]. It is imperative to accentuate that these types of dietary cholesterol and cholesterol made from scratch are necessary for intestinal integrity. Inhibitors of cholesterol physiology have long been utilized in treating and preventing cardiovascular diseases associated with hypercholesterolemia [40]. The sole obvious success in this field is Ezetimibe, and researchers are still looking for other medications with equal or comparable efficacy. In the past, concerned genetic routes have been examined and considered as a top experimental model, which can aid the exploration of the mechanism of cholesterol absorption [41]. This evidence can be applied as a crucial tool to determine better pharmacological agents. The best lipid-lowering treatments may be developed because of improved sympathetic impacts on genetic variants. An overview of cholesterol absorption and its intricate relationships with the routes can be explained by addressing a few queries, including the absorption of cholesterol from intrinsic and external sources (**Figure 6**). Although they are connected and contribute to the overall cholesterol homeostasis in the body, from a physiological perspective, cholesterol production and absorption are distinct processes. Thus, this section provides a brief overview of the research on cholesterol absorption as well as several gene candidates suggested as prospective therapeutic targets.

A different approach is also offered in addition to a thorough description of the most popular and adaptable technique for determining cholesterol absorption, along with crucial points to keep in mind when interpreting the results. Recently, drug development researchers have laid great emphasis on reverse cholesterol transport (RCT), which has progressively been understood to be another critical step in reducing the risk of CVD [43]. This study offers a detailed description on identifying neutral and acidic fecal sterols and analyzing the data. Sterol excretion is the primary

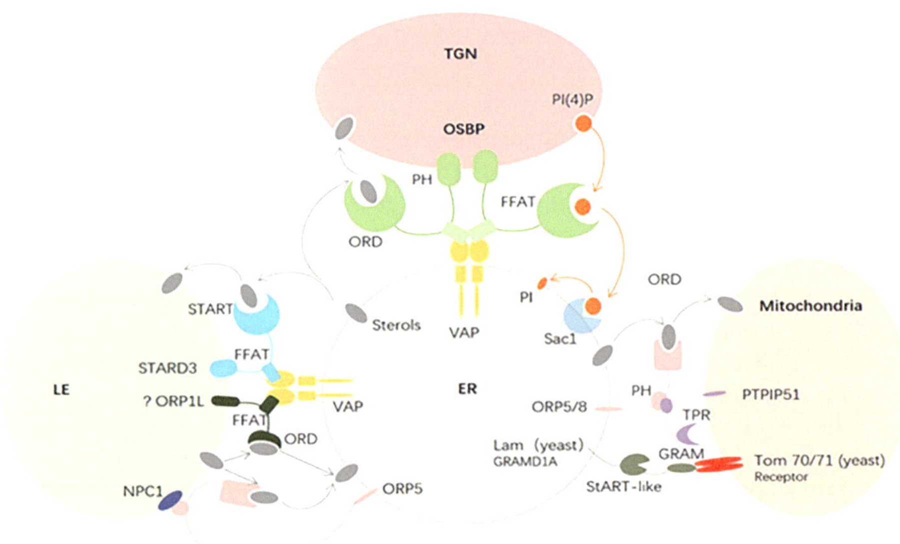

Figure 6.
An illustration of major molecules in intracellular cholesterol transport [42]. Reprinted (adapted) with permission from. Copyright (2022) Xiaochun Tang et al. Front. Cell. Dev. Biol., Frontiers.

sign of RCT [44]. It has been observed particularly in the case of gene-diet interactions, and further advancements in research will facilitate significant strides in the examination of the role of genetics on individual changeability in cholesterol retort. This includes polymorphisms of various proteins/carriers and enzymes tangled in the intestinal uptake or transfer of cholesterol.

4. Cholesterol biosynthesis and enzymatic control

Cholesterol is a crucial part of cell membranes and the building block for the production of bile acids and steroid hormones. Part of its production occurs in a membranous environment, specifically in the latter stages, where relevant enzymes, products, and substrates typically tend to have very high hydrophobicity. Over the past 50 years, the importance of cholesterol has increased due to its connection to cardiovascular diseases, one of the leading causes of death worldwide. To meet the present need for novel drugs that can efficiently control the content of cholesterol in the blood, it is imperative to comprehend how cholesterol is produced in the body and identify the major enzymes involved in the process. In this context, the features and catalytic machinery of the enzymes participated in the creation of cholesterol from its original two-carbon building block, acetyl-CoA, will be reviewed, and their present therapeutic relevance will be investigated [9]. Mechanism detects cholesterol and oxysterols closely and controls the generation and uptake of cholesterol. Acetate serves as the precursor for the more than 30 chemical processes necessary for cholesterol production, which occurs in the endoplasmic reticulum (ER). Thiolase and HMG-CoA synthase catalyze the first two reversible steps, which result in the condensation of three acetate molecules into coenzyme A (3-hydroxy-3-methylglutaryl-CoA), resulting from the condensation of three acetate molecules into acetoacetyl-CoA [45]. The key point of regulation in cholesterol synthesis is the succeeding process, which is catalyzed by the enzyme HMG-CoA reductase, a transmembrane protein of ER. HMG-CoA is reduced to mevalonate. A sequence of reactions convert mevalonate into isopentenyl pyrophosphate. Six isopentenyl pyrophosphate molecules are encouraged to condense in the following stage, resulting in squalene, which is then cycled and converted into cholesterol by a variety of mechanisms [46]. The discussion in this section will be an invaluable tool for undertaking future studies on the cholesterol synthesis pathway and in turn cultivating a deeper knowledge of cholesterol metabolism.

SREBP-2 coordinates the role of cellular cholesterol in regulating HMG-CoA reductase transcription, the key enzyme involved in cholesterol synthesis [47]. Transmembrane sterol-sensing domains (SSDs) are present in HMG-CoA reductase and exhibit strong similarity with SSD found in SREBP cleavage-activating protein (SCAP) [48]. As previously shown, this domain is crucial for the attachment of Insig-1 or Insig-2, proteins necessary for enzyme breakdown and its regulatory monitoring [49]. The inset recruits HMG-CoA reductase when there is a high concentration of cholesterol present because it is intrinsically related to the membrane-destined ubiquitin ligases presented in ER membrane. An ATPase known as vasolin-containing protein recognizes the HMG-CoA reductase, which leads to the proteasomal degradation of the reductase in a sterol-Insig-dependent manner. This recognition is made possible by the ubiquitination induced by ubiquitin ligases. Oxysterols such 24-, 27-, and 25-hydroxycholesterol can also lead to HMG-CoA reductase being ubiquitinated by Insig-1 [50]. Additionally, the binding of oxysterols to Insig-1 prevents the transport of SCAP-SREBP-2 to the Golgi and, subsequently, the translocation of the transcription factor in the nucleus [51]. To decrease cholesterol synthesis, oxysterols inhibit HMG-CoA reductase at both the mRNA and protein concentration [52]. Lanosterol and 24, 25-dehydrolanosterol intermediates in the production of cholesterol promote HMG-CoA reductase ubiquitination and destruction of Insig-reliant without impeding SREBP-2 treating [53]. Other intermediates, including 25-hydroxycholesterol, 27-hydroxycholesterol, 7-Keto, and 27-hydroxylanosterol, exhibit the similar result.

The LDL receptor alters the absorption of cholesterol, and the transcription factor sterol regulatory element-binding protein-2 regulates the levels of key enzymes such as HMG-CoA reductase, HMG-CoA synthase, and mevalonate kinase genes. The NH2- and COOH-terminal domains of SREBP-2, a double-helix transmembrane protein, are located on the cytoplasmic side and are connected to the endoplasmic reticulum membrane [54]. The active territory exists at NH_2-terminus and touches the nucleus and finally fixes to the sterol regulatory element (SRE). Following this, SREBP-2 triggers transcription of the genes it is intended to affect. SREBP has connections with insulin-induced gene protein-1 and the SREBP cleavage-activating protein, both of which have eight membrane-spanning sections of Insig-1. When cholesterol levels drop, the SCAP-SREBP complex separates from Insig-1, which is degraded by proteasome after being ubiquitinated. A small GTPase called Sar1 facilitates the process of sorting the complex into coat protein complex II or (COPII)-coated vesicles. Sec23/Sec24 and Sec13/Sec31 (Sec23-Sec24 complex (Sec23/24), a crucial cargo-binding adaptor, and Sec13-Sec31 (Sec13/31) formed structural outer layer during vesicle creation) are two examples of coat proteins that are sequentially recruited and clustered [55]. The SCAP-SREBP complex is then created and transported by COPII coat vesicles to the Golgi, where it is met with two local proteases, site-1 (S1P) and site-2 (S2P), which cleave SREBP and liberate the NH2-active domain (**Figure** 7) [56].

SREBP-2 functions as a regulatory mechanism that not only promotes the expression of genes involved in cholesterol production and absorption, but also has the ability to bind to the E-box in the promoter region of the ATP-binding cassette

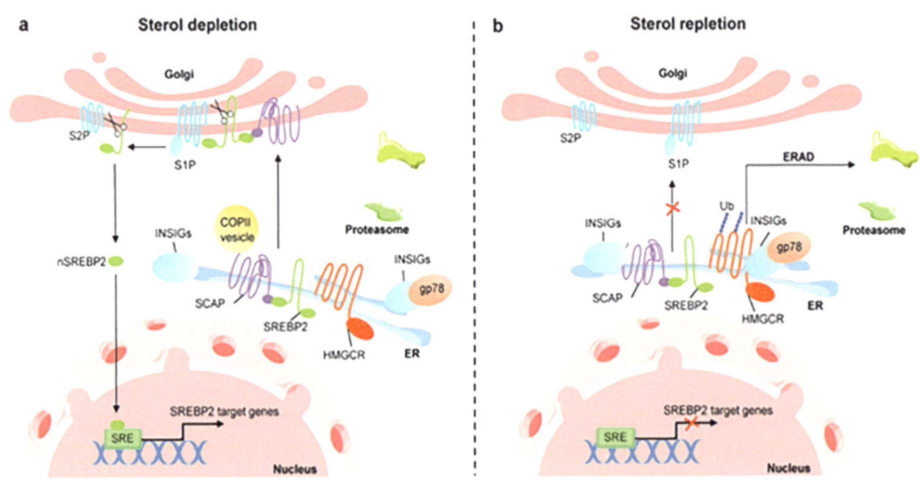

Figure 7.
SREBP2 pathway in cholesterol production regulation. The sterol regulatory element requisite protein paths and the HMG-CoA reductase. The degradation pathway represents negative feedback regulation, which carefully controls the process of cholesterol biosynthesis [12]. Reprinted (adapted) with permission from. Copyright (2022) Jihong Han et al. Signal Transduction and Targeted Therapy, Nature.

transporter (ABCA1), preventing cholesterol efflux by blocking its transcription. Additionally, SREBF2 co-transcribes with miR-33, a microRNA that prevents cholesterol export and trafficking to quickly raise intracellular cholesterol concentrations. Located at the SREBF2 gene is miR-33 [57]. Because SREBP cleavage-activating protein (SCAP) undergoes a number of conformational changes that make it easier for it to bind to Insulin-induced gene protein-1, the Insig1-SCAP-SREBP2 complex is bound in the ER when there is a high level of cholesterol (Insig-1). This behavior is thought to be caused by the sterol-sensing domain (SSD) in SCAP, which functions as the cholesterol-dependent active site for Insig-1. SCAP stops Insig-1 from being ubiquitinated and destroyed, stabilizing it at the protein level [58]. In the presence of cholesterol or 25-hydroxycholesterol in the ER membrane, coated proteins are unable to bind to SCAP, preventing the formation of COPII-coated vesicles [59]. This procedure is managed by the SCAP Loop 6 sequence methionine-glutamate-leucine-alanine-aspartate-leucine (MELADL). The migration of SCAP-SREBP from the ER to the Golgi has been demonstrated to depend on the presence of a number of domains in the SCAP sequence [60]. The hexapeptide sequence, MELADL sequence at Loop 6, has a place where coat proteins can bind [61]. As a result, cholesterol loading is not possible because COPII-coated vesicles cannot be assembled because the conformational change brought on by Insig-1 binding, which moves the MELADL sequence and makes the binding site inaccessible, precludes it.

Cholesterol binding to luminal loop 1 causes SCAP to lose its connection to loop 7 and adopt an open conformation that makes it easier for Insig-1 to bind to SCAP and prevents COPII proteins from reaching the MELADL sequence, keeping the SCAP-SREBP-2 complex imprisoned in the ER. It has recently been demonstrated that the ubiquitin E3 ligase Rnf145, Ring finger protein 145, can also suppress the transcriptional activity of SREBP2 [62]. This LXR target gene causes cleavage-activating protein (SCAP) ubiquitination on two lysine residues near the COPII-vesicle binding site, which prevents SCAP-SREBP-2 from being transported to the Golgi. This demonstrates how sterol regulatory element-binding protein (SREBP) and LXR, the dominant watchdogs of cholesterol creation and efflux, interact to keep cholesterol levels imbalance (**Figure 8**). Another point of control on the synthesis of cholesterol is catalyzed by the enzyme squalene monooxygenase (SM), which is involved in the conversion of squalene into the precursor of lanosterol known as 2,3(S)-mono-oxidosqualene (MOS) [64]. Since SM is intimately connected to the ER membrane and SREBP-2 also regulates its transcription, it follows that the stability of cell depends on the concentration of cholesterol. Studies conducted in vitro revealed that the proteasomal system of cholesterol, nonetheless oxysterols, caused SM polyubiquitination and deterioration. Both RING finger 6, an E3 ubiquitin ligase, was responsible for mediating this pathway, which was Insig-independent (MARCH6) [65]. It's interesting to note that the N-terminus of SM contributes to the conformational changes in SM's structure brought on by cholesterol, which are necessary for SM's degradation that is dependent on cholesterol. The HMG-CoA reductase activity is specifically inhibited by statins, a class of drugs used to lower plasma cholesterol levels. This stops the body from producing cholesterol.

To increase the consumption of LDL-c particles and lower plasma cholesterol levels, LDLR is consequently increased in hepatocytes. There were reductions in major vascular events and all-cause mortality among persons without symptoms of cardiovascular disease, according to a meta-analysis from the Cochrane Library for the use of statins in primary prevention. Statins have long been known to decrease LDL-c plasma extent in patients with erstwhile cardiovascular diseases.

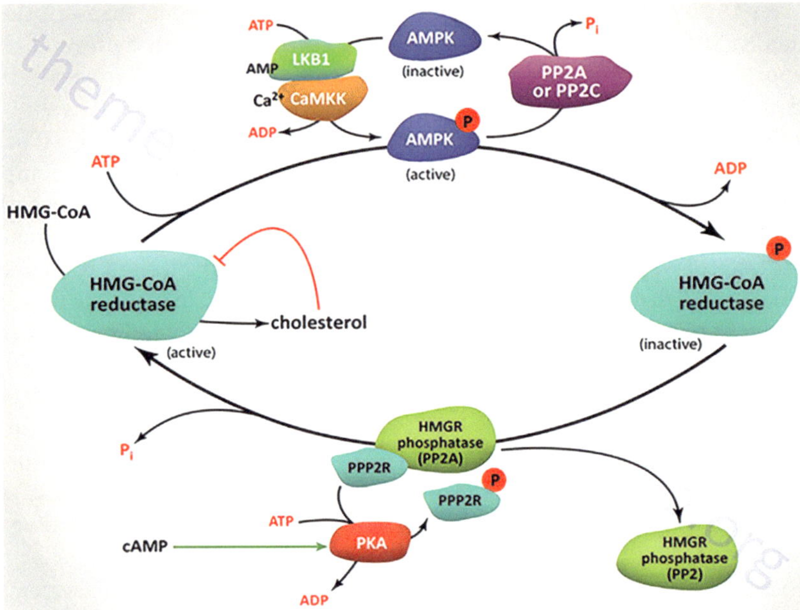

(a)

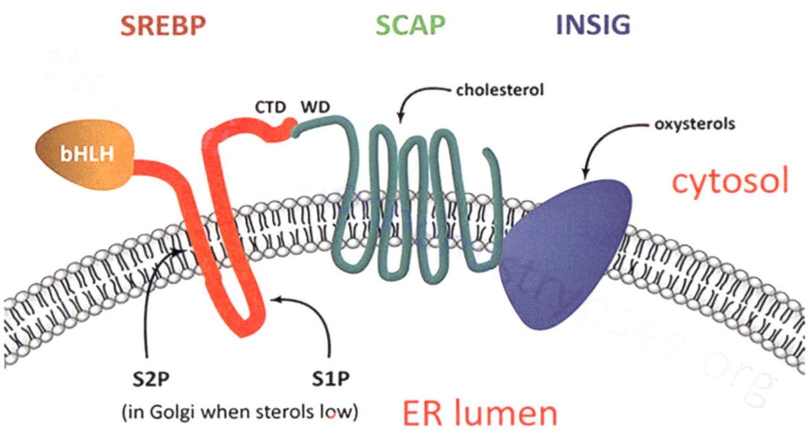

(b)

Figure 8.
(a) Regulation of HMGR by covalent modification. (b) Protease-mediated regulation of SREBP activation [46]. Reprinted (adapted) with permission from [63]. Copyright © 1996–2022 the medical biochemistry page, LLC.

5. Cholesterol balance

Low cholesterol has been linked statistically to greater rates of violent behavior, Parkinson's disease cancer, and suicide mortality. It has also been linked to bipolar illness, depression, and anxiety. Lower cholesterol levels are also related to gastro-intestinal illnesses and TB susceptibilities. Every cell of an animal has cholesterol, a sterol that is necessary for life. Our cells' essential structure includes cholesterol, which safeguards our tissues. Many individuals dread and concentrate on having high cholesterol levels, but they never pay attention to how low cholesterol levels can affect their health. Balance is vital, as it is in everything else. Statistically speaking, a higher level of cholesterol is accompanying with a high percentage of threat of cardiovascular illness; but low cholesterol is linked to poor health and various chronic diseases, which is a less well-known association. An inferior concentration of cholesterol does not necessarily translate to a longer lifespan or higher quality of life, according to recent studies [66]. A high cholesterol level or a cholesterol deficit can be identified using the reported profile. Additionally, these types of profiles identify risk factors for neurological and/or vascular disease and assess the body's elimination of potentially harmful homocysteine. Maintenance of a healthy balance of cholesterol throughout the body depends on the close link between cholesterol production and absorption. While synthesis increases, absorption declines, or vice versa, because of compensatory mechanisms shifting the routes in opposite directions. The influence of atorvastatin involved overexpression of genes for cholesterol production and absorption of SREBP-2, Niemann-Pick C1-Like 1 (NPC1L1), HMG-CoA reductase [67], and LDLR) and downregulation of genes for cholesterol efflux from enterocytes back into the intestinal lumen (ABCG5 and ABCG8) [35]. The balance between dietary cholesterol, cholesterol excretion, and endogenous synthesis is supported by this compensatory system. Endogenous reaction to the rise in dietary cholesterol was explored in several studies employing sterol balancing techniques, and it was shown that participant responses varied greatly. Variations in the transcriptional and posttranslational mechanism of proteins involved in cholesterol uptake, efflux, and production are responsible for this variability. Increased bile acid production that is expelled in the stool, the primary method for excreting cholesterol, balances out the rise in dietary cholesterol. It was reported earlier that an enhancement in cholesterol consumption decreased cholesterol production and increased endogenous cholesterol evacuation through the biliary tract [68]. Thus, cholesterol production should be decreased in order to have a powerful compensatory mechanism that stops the rise in plasma total cholesterol that results from an enhancement in cholesterol consumption.

Due to a reduction in exogenous cholesterol absorption, an increase in dietary cholesterol has little impact on the concentrations of entire plasma and LDL cholesterol. When dietary cholesterol was increased from 240 to 800 mg per day utilizing radiolabeled acetate integrated into sterols on outlying blood mononuclear cells, endogenous cholesterol production was estimated to have decreased by 21% [26]. However, individuals who did not reduce cholesterol production to offset the increase in cholesterol consumption experienced a considerable rise in plasma cholesterol concentration. Dietary or endogenous sources are where the intestinal lumen's supply of cholesterol comes from (intestinal epithelial sloughing, bile secretion, and TICE). Liver transforms cholesterol into bile acids, which are then coupled with either taurine or glycine and released into the bile before being further removed in the stool (**Figure 9**) [69]. Bile acid loss in the feces (5%), which is compensated for by the liver's tightly controlled synthesis of bile acids, which is depressingly impacted by bile acids reabsorbed, occurs [70]. Every type of mammalian cell contains cholesterol, which is a subtype of lipid.

It serves as a crucial component of cell membranes by preserving their integrity and stability as well as a precursor to many forms of essential sterol compounds, such as vitamins and hormones. Similar to other significant substances, the metabolism controls the amount of free cholesterol in the body by de novo production, intake, export, and esterification. More than 20 enzymatic platforms make up the de novo cholesterol production, also known as the mevalonate route, in various circumstances. Three acetyl-CoA molecules are combined to create one 3-hydroxy-3-methylglutaryl coenzyme A as the initial step (HMG-CoA). HMG-CoA reductase (HMGCR) [71],

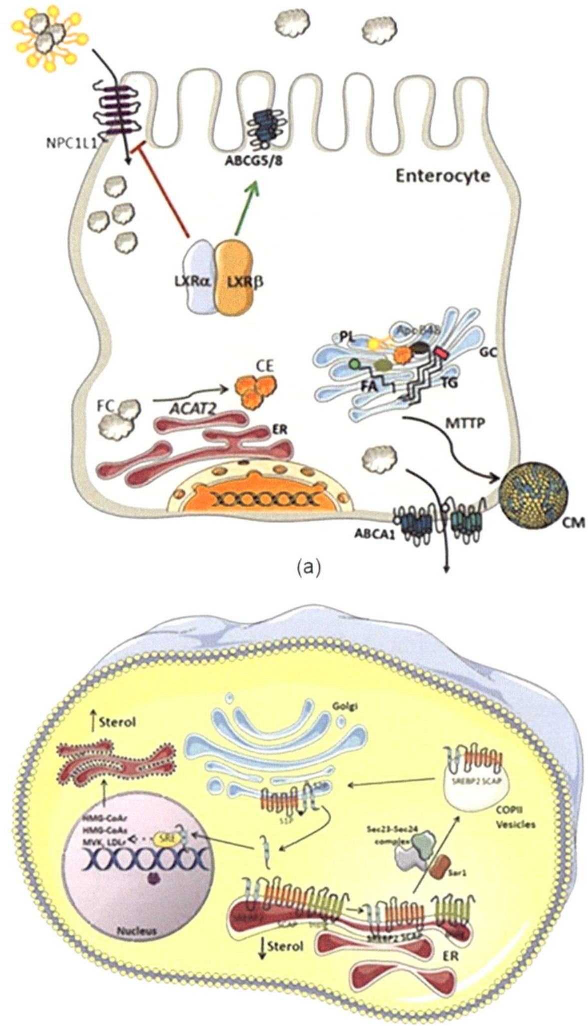

Figure 9.
(a) Mechanisms for absorbing cholesterol. Through NPC1L1 and ACAT2 in the endoplasmic reticulum, free cholesterol (FC) carried in micelles enters the enterocyte (ER) [26]. (b) Regulating the manufacture of cholesterol. SCAP and Insig-1 work together to attach SREBP-2, the principal regulator of cholesterol production and absorption, to the ER membrane [26]. Reprinted (adapted) with permission from. Copyright (2018) Milessa Silva Afonso et al. Nutrients, MDPI.

the first rate-limiting catalytic enzyme, changes HMG-CoA into mevalonate, which then undergoes a sequence of enzymatic processes to become farnesyl pyrophosphate (FPP), squalene, and ultimately cholesterol [72]. Notably, FPP can transform into geranylgeranyl pyrophosphate, which is another crucial factor in protein prenylation, in addition to downstream sterols and all other nonsterol isoprenoids.

MicroRNAs (miRNAs) can posttranscriptionally control gene expression [73]. Numerous miRNAs in human being participate in controlling practically every function, including transport, synthesis, and upkeep of cholesterol homeostasis. MiRNAs are appealing targets for monitoring dyslipidemias and supplementary lipid-interlined with illnesses because of their tiny size and capacity to extremely precisely govern gene expression. Because of the intricate relationships between miRNAs, transcription factors, and gene expression, there is a high risk that miRNA overexpression or reticence will have negative effects [74]. Numerous food ingredients have displayed potential in targeting particular miRNAs and changing the appearance and expression of genes downstream. Therefore, much more investigations must properly comprehend the function(s) of each miRNA in the body and how nutrition and health may affect those functions. This chapter gives a general overview of the roles that miRNAs are known to play in controlling reverse cholesterol transport, maintaining cholesterol homeostasis, and potential clinical repercussions of their change [75]. This conversation will be profoundly engrossed in the role of kinases in cholesterol efflux, which has been the theme of the mainstream of research to date (particularly on the ATP-binding cassette transporter A1).

6. Cholesterol efflux

An initial step in this mechanism, cholesterol efflux from cholesterol-rich macrophages, has been demonstrated in animal models to prevent atherosclerosis. Researchers examined the relationships between cholesterol efflux prospective and cardiovascular death and various common cardiovascular biomarkers and risk variables to get a further understanding of the regulation of cholesterol efflux [76]. Cells must keep their cholesterol levels within very strict boundaries since excess or fewer amounts of cholesterol might disturb apoptosis, necrosis, and cellular membranes. Both intracellular production and plasma lipoproteins can supply the cholesterol needed by cells, and both sources are adequate to meet those needs. Cholesterol deficiency is uncommon because the mechanisms of making and absorbing cholesterol are strictly controlled. Excessive cholesterol is the most common problem [77]. Hepatocytes and, to a lesser extent, adrenocortical cells are the only cells that can break down cholesterol in the body. To lower their cholesterol levels, cells can either efflux cholesterol, which have potentially limitless capacity or convert cholesterol into cholesteryl esters, which have a limited capacity because it is hazardous to over-supply cells with cholesteryl esters. This distinct method of cholesterol efflux is regulated by a multitude of intracellular transporters, including scavenger receptor type B-1, and ATP-binding cassette transporter proteins. The levels of fibrinogen, C-reactive protein, interleukin-6, and serum amyloid A were all adversely correlated with cholesterol efflux [78]. Patients with cholesterol efflux levels in the lowest quartile had a greater rate of cardiovascular death. Apolipoprotein A-I and high-density lipoprotein are natural plasma acceptors of cholesterol. HDL-related markers, such as HDL cholesterol, apolipoproteins AI, AII, HDL phospholipids, and HDL particle amount, showed a substantial correlation with cholesterol efflux, whereas LDL-related markers, such as LDL cholesterol and apoB, did not [79]. Once HDL cholesterol was considered, this connection was lessened but not eliminated (**Figure 10**) [81]. Researchers displayed that the cholesterol efflux was related to HDL composition and inflammatory burden and that this negatively predicts cardiovascular mortality irrespective of HDL cholesterol during coronary angiography. This section provides the information on: (I) how a specific treatment (a mutation, an overexpression or a therapy, a knockdown) affects a cell's ability to efflux cholesterol; and (II) how a disease or a treatment affects the ability of plasma acceptors to take up cholesterol. This method is often used in the context of cardiovascular research, infectious reproductive diseases, and metabolic and neurodegenerative disorders. High-density lipoprotein (HDL) plasma levels are negatively correlated with the prevalence of cardiovascular illnesses.

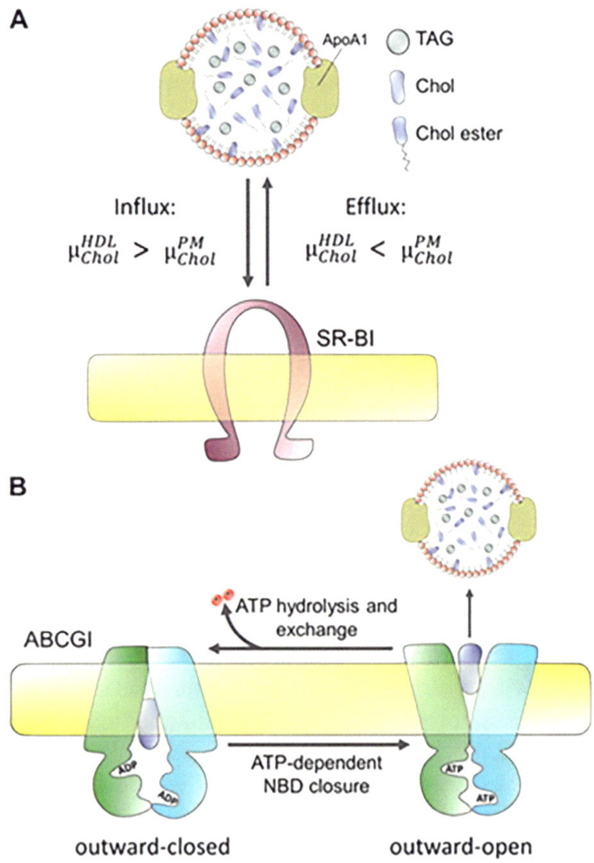

Figure 10.
Cholesterol efflux from mammalian cells is protein-mediated both passively and actively. Reprinted (adapted) with permission from [80]. Copyright (2022) Daniel Wüstne et. al. Front. Cell Dev. Biol., Frontiers.

Mendelian randomization analysis and various clinical experiments are not finding any importance of plasma HDL-cholesterol enriching medications on cardiovascular illnesses and have cast doubt on the causal association between the concentration of plasma HDL-cholesterol and cardiovascular illnesses [82]. However, underlying connection between the concentration of HDL-cholesterol and cardiovascular health has been established recently. Mendelian randomization trials showed that most cases of cardiovascular disease proved the association of HDL-cholesterol and cardiovascular health in these investigations. As well, numerous studies in sizable population cohorts have demonstrated an inverse correlation between cardiovascular diseases and HDL's ability to remove cholesterol from the body.

The cholesterol efflux pathways exhibit anti-inflammatory and anti-atherogenic properties. The phenomenon of cholesterol efflux controls the rate of proliferation of hematopoietic stem and predecessor cells. This process governs the physiology of inflammation and Inflammasomes during the activation of the macrophages. The accumulation of cholesteryl esters in macrophages, or the development of macrophage foam cells, is likewise inhibited by cholesterol efflux channels [83]. Recent research on single cells using RNASeq showed that atherosclerotic plaques have existed in macrophage foam and that these cells also contained fewer pro-inflammatory genes than non-foam cells [84]. This only occurred because of stimulated liver X receptor, G1 cholesterol transporters, increased ATP-binding Cassette A1 reactivity, and reduced inflammation. However, anytime such pathways are blocked, effected cells turn into inflammatory foam.

7. Cell signaling and cholesterol homeostasis

Complex systems involving synthesis, export, and uptake are used to maintain cholesterol homeostasis. To explore the phenomenon of the participation of signaling in these routes, a lot of research has been done to clarify these cholesterol-related pathways. All mammalian cells require cholesterol to function properly. To meet cellular cholesterol needs, plasma lipoproteins are synthesized from scratch and taken up by the cells in a process known as homeostasis, which is transcriptionally controlled by proteins known as sterol regulatory element-binding proteins (SREBPs) [85]. Together with other members of the nuclear receptor superfamily, farnesoid X receptor (FXR) and liver X receptors (LXRs) enhance catabolism, storage, and transit of sterols as well as their metabolites to lessen the cytotoxicity brought on by accumulating excessive amounts of cholesterol. These metabolic nuclear receptors contain LXR, PXR, CAR, and FXR, as well as bile acid, fatty acid, and oxysterol receptors [86].

Through the coordinated regulation of transcriptional programs, these nuclear receptors regulate crucial processes including inverse cholesterol transport and absorption, bile acid formation, lipoprotein uptake, lipoprotein synthesis, and reconfiguration by peripheral tissues. This chapter reviews various routes of signaling that affect abovementioned processes and rate of cholesterol homeostasis, as well as transcriptional events. This section's main objective is the involvement of kinases in cholesterol efflux, which has been the subject of most research to date, particularly on the role of signaling in cellular cholesterol homeostasis and the ATP-binding cassette transporter [87]. This update emphasizes the important role of nuclear receptors in coordinating the intricate transcriptional programs that control cholesterol metabolism and the synthesis of related bile acids, the interactions of lipid metabolites with their receptors, and the nuclear receptors' function in lipid signaling cascades. It's been known for a long time that cholesterol serves various purposes in mammalian cells. Cholesterol is essential for the production of steroid hormones, bile acid metabolism, and cellular structures such lipid rafts in addition to playing a key role in cell membranes. Because cholesterol plays so many different and important roles in cells, its levels must be regulated within a certain range. Mammalian cells include complex cholesterol homeostatic machinery with multiple levels of regulation. The cholesterol synthesis pathway and immersion through low-density lipoprotein receptor are the two best-understood ways for most of the cells to obtain cholesterol [11]. Cholesterol can be taken up by some cells by phagocytosis and exchange routes, including macrophages. The majority of cell types have a cholesterol export route, which becomes active when cellular cholesterol concentration is excessive and needs to be reduced. Many common disorders, including atherosclerosis, where cellular cholesterol accumulation impairs regular cellular functioning, are caused by an imbalance between uptakes, synthesis, and export.

To identify new treatment targets, significant research is still being conducted to understand the precise controlling of cholesterol homeostasis [88]. The significance of posttranslational mechanisms concerning cholesterol homeostasis is growing along with transcriptional regulation. Cell signaling pathways include a posttranslational switch, which is a rapid way to turn proteins on and off before transcriptional mechanisms have a time to take effect. With the creation of multiple effective medications that target cell signaling networks, it is now more crucial than ever to understand the function of these pathways. Since much of the research to date has concentrated on this topic, the authors of this chapter intend to highlight some significant advancements in the function of signaling routes controlling cholesterol concentration in cells. Signaling in cholesterol efflux will receive special attention [89]. Complex protein interactions that entail regulating protein phosphorylation states by kinases and phosphatases are a part of cell signaling cascades [80]. Many signaling pathway intermediates and components interact with one another; therefore, it might be difficult to pinpoint the precise kinase or phosphatase responsible for an observed impact. This is a warning that applies to many studies in this field. Transcriptional regulation is the level of cholesterol homeostasis that has been the most thoroughly studied. In terms of important transcriptional regulators that have been found to be necessary for the creation of cholesterol and the absorption of LDL, the transcription factor SREBP-2 is at the top of the list. SREBP-2 is proteolytically activated when endoplasmic reticulum cholesterol levels are low, which causes the transcription of genes involved in production and uptake of cholesterol and 3-hydroxy-3-methylglutaryl coenzyme A reductase to be induced low-density lipoprotein receptor [90]. A network of transcriptional processes is responsible for maintaining cellular and overall body cholesterol homeostasis (**Figure 11**). A negative feedback loop that responds to increases in intracellular cholesterol at various levels precisely controls de novo cholesterol synthesis and lipoprotein cholesterol absorption in cells.

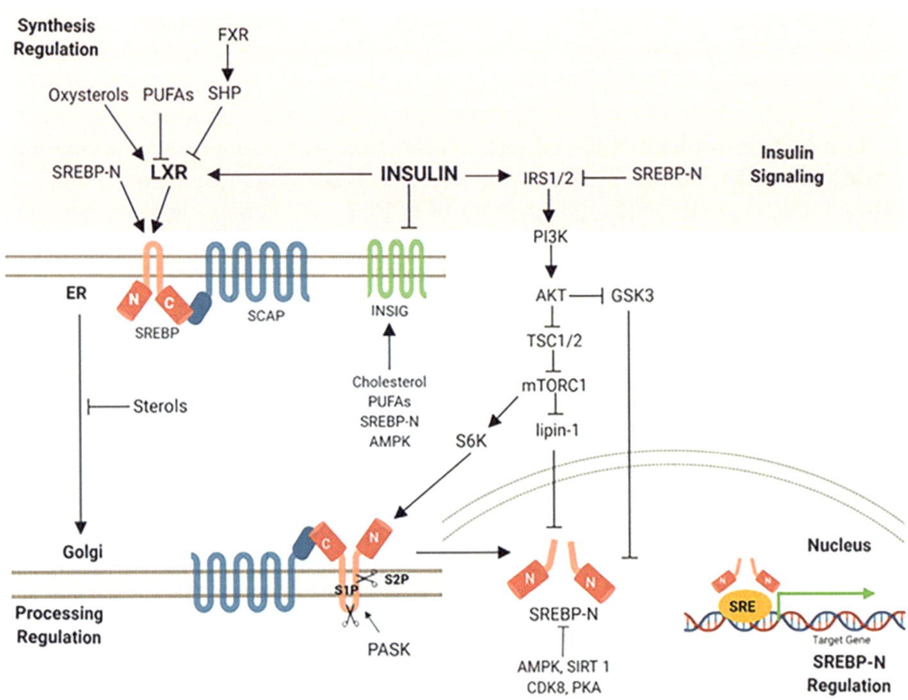

Figure 11.
Regulation of SREBPs occurs at various stages including, multiple signals regulate SREBP synthesis, transcriptional activity, degradation, and proteolytic activation. Activated, nuclear SREBP is tightly controlled by posttranslational modifications. Reprinted (adapted) with permission from [91]. Copyright (2020) Hunjoo Haet al. Front. Pharmacol. Frontiers.

The membrane-bound transcription factors (SREBPs) directly stimulate the features of genes participated in cholesterol production and uptake as well as lipogenesis, controlling this pathway [92]. The SREBP pathway ensures that there is enough cholesterol for cellular needs, but excessive amounts of free or surpass cholesterol because the production of SREBP-dependent genes to be suppressed (**Figure 12**). Along with other members of the liver X receptors, family of nuclear receptor and

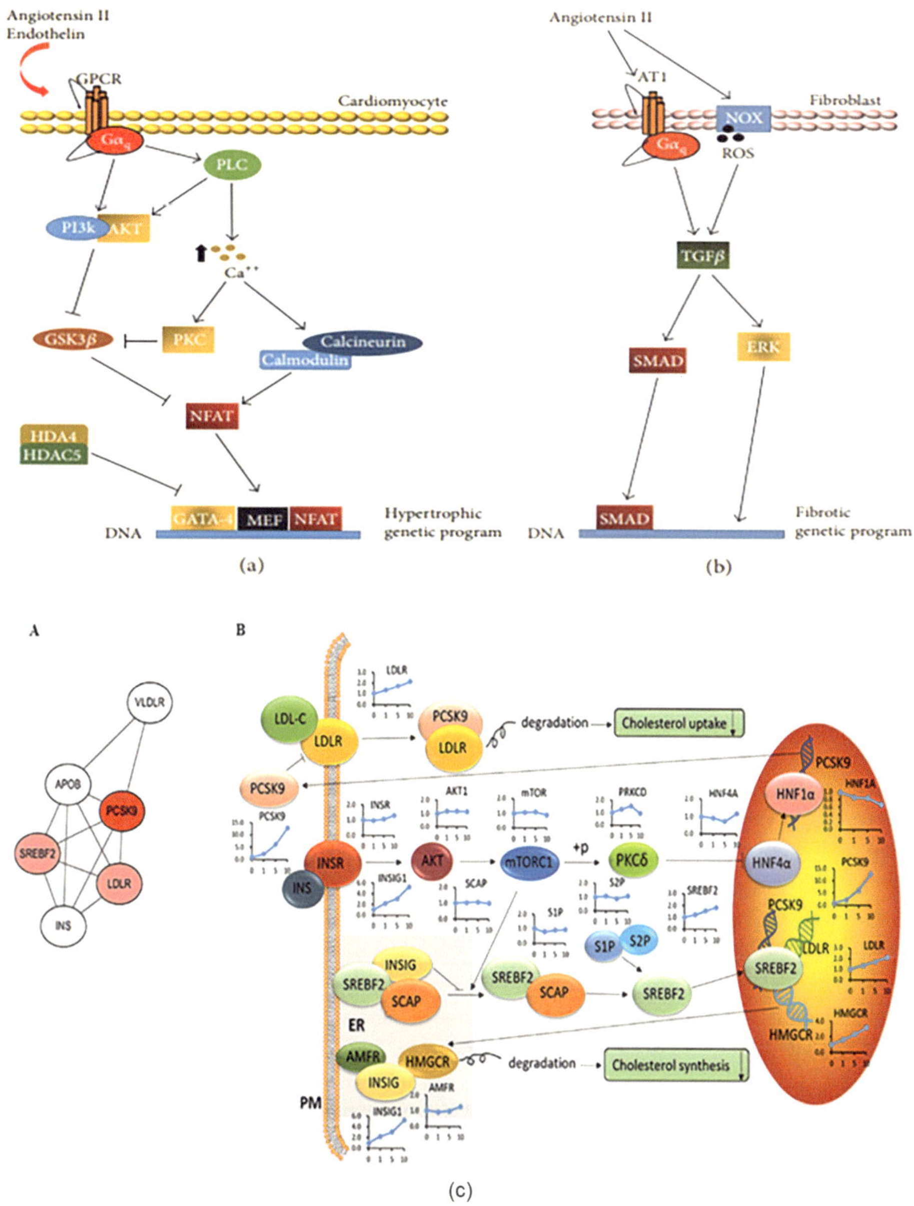

Figure 12.
(a) and (b) Signaling pathways for myocardial fibrosis and hypertrophy (a, b). Several substances take role in the control of the genes responsible for cardiac hypertrophy. Reprinted (adapted) with permission from [15]. Copyright (2017) Valentina Valenti et al. Oxidative Medicine and Cellular Longevity, Hindawi. (c) A description of the main signaling pathways involved in the metabolism of cholesterol. Reprinted (adapted) with permission from [93]. Copyright (2017) Mingyan He et al. Oncotarget.

the farnesoid X receptor facilitate the storage, transit, and catabolism of sterols to prevent cholesterol buildup [94]. The bidirectional movement of cholesterol between peripheral tissues and liver and the hepatic elimination dietary sterols, sterol metabolites, and of cholesterol are all mediated by these metabolic nuclear receptors. LXR (NR1H3), LXRs, and LXR (NR1H2) are nuclear receptor superfamily members and ligand-activated transcription factors.

To trigger gene expression, LXRs preferentially attach to LXR response elements, which consist of two hexanucleotide repeats disconnected by four nucleotides, together with their heterodimeric partner, and retinoid X receptor (RXR) [95]. With lower expression in the splenic tissues, adipose, gut, and kidney, LXR abundantly occurred in the liver tissue. Recent research has demonstrated that oxysterols are a particular ligand for the LXRs. The most powerful oxysterols are 22(R)-hydroxycholesterol, an intermediary in the synthesis of steroid hormones, 24(S),25-epoxycholesterol, which is formed in 24(S)-hydroxycholesterol, termed as hepatocytes and macrophages, are common cholesterol metabolite in brain tissue [95]. Moreover, the 27-hydroxycholesterol formed in macrophages to counter cholesterol loading and activate LXRs is a significant physiological ligand. Transactivation of the genes ABCA1, APOE, CETP, and PLTP, as well as ABCG1, ABCG5, and ABCG8 involved in sterol transport, cholesterol efflux, sterol catabolism, and high-density lipoprotein metabolism is how LXRs react to high cholesterol levels (CYP7A1) [96]. LXRs also play a significant role in regulating the number of lipids present in cells by activating SREBP-1c, the main regulator of de novo lipogenesis. Two examples of the fatty acid synthesis genes that SREBP-1c regulates are stearyl-CoA desaturase-1 and fatty acid synthase. Stearyl-CoA desaturase-1, on the other hand, transforms stearyl-CoA into oleyl-CoA, the preferred substrate for cholesterol acyltransferase, which uses acyl-CoA [97]. Combining the control of de novo lipogenesis with cholesterol catabolism allows LXRs to remove extra sterol more effectively and stop cholesterol-induced cytotoxicity. SREBP-1c-mediated fatty acid synthesis coped with additional free cholesterol when there was high cholesterol. The presence of additional cholesterol in cells serves as a temporary buffer for free cholesterol that is present at the cellular level because the CE cycle continuously converts the cholesterol associated with lipid droplets. However, the production of LXR-mediated CE and triglyceride in tissues that synthesize lipoproteins provides the fundamental components for lipoprotein assemblage and secretion, serving as a significant route for sterol export [98]. LXRs performed as sterol sensors and trigger cholesterol catabolism, efflux, and de novo lipogenesis simultaneously to control the concentration of cellular cholesterol and inhibit lipotoxicity.

8. Progression of cardiac remodeling

The advancement of heart failure is connected with cardiac remodeling, which is defined as a collection of interstitial, molecular, and cellular alterations that modify the geometry of the heart and ultimately affect its size, functional style, mass, and form. Cardiac remodeling occurs throughout both the adaptive and maladaptive developmental stages [17]. When it occurs late in the process, heart failure occurs. This is an adaptive response to retain cardiac output in its initial stages. Oxidative stress appears to be the main mechanism that initiates the transformation from cardiac hypertrophy to heart attacks because of modifications in signal transduction, poor calcium handling, increasing cardiomyocyte loss, malfunction of the sarcolemma and sarcoplasmic reticulum, and increases in cardiac fibrosis [99]. In this part, the principles and medical effects of cardiac remodeling are reviewed along with the pathophysiological relevance of a number of factors, including oxidative stress, inflammation, cell death, energy metabolism, collagen, contractile proteins, and form. The study concludes by describing three different levels of techniques for the pharmacological treatment of cardiac remodeling: solidified, promising, and prospective strategies. Increases in left ventricular mass and volume have been linked directly to future left ventricular performance decline and a less favorable clinical course, according to natural history studies in heart failure [100]. Although cardiac remodeling is considered important in heart failure, little is understood about the fundamental mechanisms that cause cardiac remodeling. Recent clinical and experimental investigations that emphasize on the significance of remodeling process during heart failure are summarized [101].

This chapter describes the mechanisms that influence LV remodeling at cellular, myocardial, and chamber levels. Norepinephrine and Angiotensin II level increases demonstrate neurohormonal activation, which is a crucial element in the early phases of cardiac remodeling and is associated with ventricular hypertrophy. An important factor in the early stages of cardiac remodeling and a link to cardiac hypertrophy is neurohormonal initiation, as evidenced by the increase in norepinephrine and angiotensin II levels. Besides it, heart-failure-related late cardiac remodeling may be accompanied by protracted neurohormonal activation, inflammatory signaling brought on by elevated levels of TNF- and TGF-, and cardiac remodeling [102]. Heart failure can arise because of cardiac remodeling, which at first seems to function as an adaptive mechanism. However, as the process progresses, molecular and cellular abnormalities are linked to it, which cause heart failure. Myocardial remodeling caused by pathological molecular pathways alters the heart's current structure and results in cardiac dysfunction. Various mechanisms include myocyte loss, modification of the event of extracellular matrix homeostasis, metabolic problems, mitochondrial dysfunction, fibrosis, and faulty autophagy within the complex signaling network identified in myocardial remodeling and cardiac hypertrophy [103]. Numerous pathophysiological stressors, including pressure and volume overload, start the remodeling cascade, which first offers heart protection as a coping mechanism. But following a myocardial infarction, persistent inflammation also triggers cardiac remodeling, which, if left untreated, progresses to heart failure [104]. Here, the authors discuss the molecular mechanisms underlying cardiac remodeling with a focus on the myocardial-infarction-related pathways. A deeper understanding of the complicated cell signaling involved in cardiac remodeling may enable the development of novel therapeutic approaches to treat heart failure and the mitigation of its effects. The authors will also discuss the evidence from gene therapy techniques for regulating crucial mediators of cardiac remodeling. Development of innovative therapeutic approaches to cure heart failure and the mitigation of cardiac consequences supported by an enhanced knowledge of the cell signaling play a part in cardiac remodeling. The authors also provide evidence from gene therapy techniques for regulating crucial mediators of cardiac remodeling [105]. Development of heart failure seems to be a widespread and well-coordinated response to cardiac malfunction and injury [105].

A better knowledge of these mechanisms is essential for the improvement of cardiovascular biology and subsequent creation of focused, efficient treatment methods for patients with congestive heart failure, according to the present mechanistic viewpoint predictably shared, highly controlled events highlight the multifaceted heart failure route. Heart failure progression is accompanied by cardiac remodeling, which is the reorganization and remodeling of the heart [106]. Regardless of the cause of CHF, cardiac remodeling plays a significant role in how the condition develops clinically. Based on reliable evidence demonstrating unique changes in cell size, ability to contract, and shape during heart failure, the traditional ideas of cellular remodeling are modified. The purpose of programmed cell death and the intriguing possibility of cardiomyocyte regeneration are both the subjects of extensive research.

Notably, the growing evidence of cardiomyocyte renewal and regeneration points to the complexity and dynamic nature of cellular remodeling still remains poorly understood [106]. With the regulation of cell death and stimulation of cell renewal as a therapeutic target, it is now conceivable to develop novel therapies to regenerate and repair failing myocardium [107]. According to a systematic assessment of the currently available research, the traditional ideas and approaches for defining the control of remodeling are completely inadequate. Failure, neurohormonal remodeling, and the novel cytokine hypothesis cannot adequately explain all cellular and molecular alterations that result in heart failure. It is essential to further discover novel biomolecules and mechanisms for the coordinated control of cardiac remodeling to find new therapeutic targets and manage the remodeling process. Hemodynamic overload, neurohormonal activation, inflammation (myocarditis), and ischemia (myocardial infarction) are the causes of cardiac remodeling. Cardiac remodeling is supposed to be both a maladaptive and an adaptive occurrence [108]. Cardiovascular remodeling helps the heart sustaining cardiac output during early stages of myocardial stress. The prolonged stressful input that is an ongoing process results in progressive decompensation. After happening of these events, the heart experiences cellular alterations, including myocyte hypertrophy, increased fibrillar collagen, fibrosis, and necrosis, apoptosis, and fibroblast proliferation. It appears as changes in the heart's geometry (the chambers change from an elliptical to a spherical shape) at the macroscopic level, which is linked to progressive left ventricular failure. Additionally, this process includes irregularities in energy consumption, altered contractile protein expression or function, abnormalities in excitation-shrinkage pairing events, and variations in the extracellular matrix.

Transforming growth factor-beta is essential for the regulation of fibroblast gene expression and phenotype, which results in interstitial fibrosis. By preventing MMP expression and stimulating the synthesis of TIMPs, TGF slows down the breakdown of the extracellular matrix. Transforming growth factor-beta also triggers the transformation of certain fibroblasts into cardiac fibroblasts and promotes the creation of extracellular matrix proteins. Results of numerous researches support the notion that the RAS and transforming growth factor-beta pathways are directly related to their role in the functioning downstream of angiotensin II [109]. The key to cardiac remodeling disease may be extracellular matrix remodeling [103]. Myocardial cells and blood vessel connections are disrupted and disorganized when extracellular matrix network structure is damaged, which reduces heart function and damages structural integrity. Increased stiffness of myocardial wall, systolic and diastolic dysfunction, and deformed architecture are caused by fibrosis and overproduction of extracellular matrix proteins. Another goal of this chapter is to outline the mechanism of cardiac remodeling as it occurs at the onset of heart failure. Cardiac fibroblasts make up about two-thirds of the heart's cell population and are the size of a cell population. Alternatively, roughly two-thirds of the cardiac tissue is made up of cardiomyocytes. This in addition to being crucial for sustaining cardiac geometry, assembly, and metabolic processes [110]. Cardiac fibroblasts are crucial for the best possible electrical conduction in the myocardium. Cardiac fibroblasts are crucial to homeostasis and remodeling of extracellular matrix (**Figure 13**).

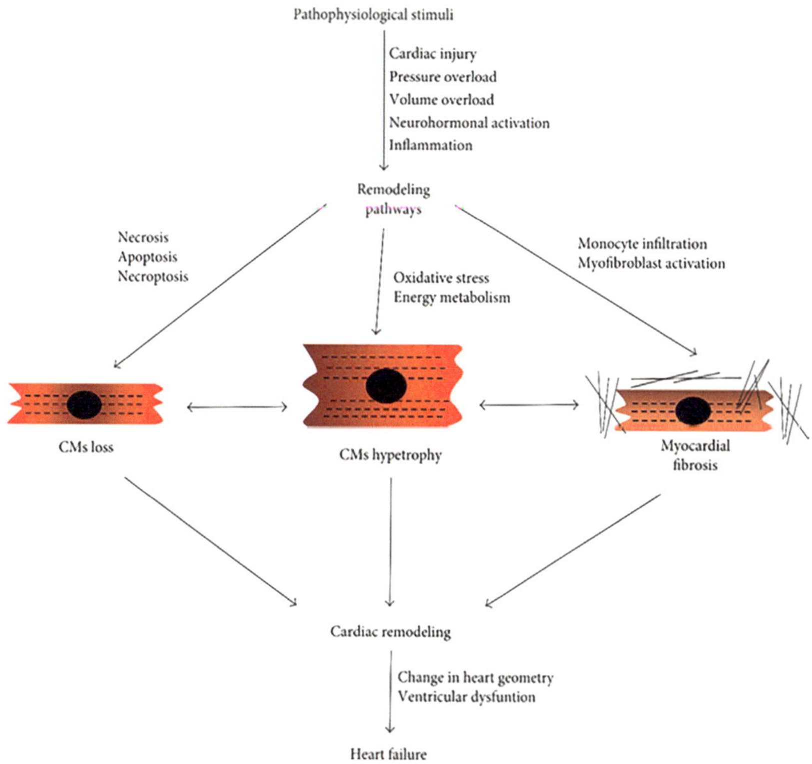

Figure 13.
Overview of the key processes leading to cardiac remodeling in a diagram. Cardiomyocyte (CM) loss, hypertophy, and myocardial fibrosis are caused by the convergence of the numerous signaling pathways involved [7], including the rise and oxidative stress pathways, in cell death, inflammation, and changes in energy metabolism. These changes result in cardiac remodeling [15]. Reprinted (adapted) with permission from [15]. Copyright (2017) Valentina Valenti et al. Oxidative Medicine and Cellular Longevity, Hindawi.

Fibronectin, elastin, glycoproteins, laminin, fibrillin, and fibrillar collagen types (I) and (III), and proteoglycans make up the ECM under normal circumstances; cardiac fibroblasts are the main source of these extracellular matrix proteins [111]. Cardiac fibroblasts also create extracellular-matrix-monitoring proteins called matrix metalloproteinases and tissue inhibitors of MMPs [112]. Extracellular matrix homeostasis depends on MMPs and proteases that break down extracellular matrix proteins, and tissue inhibitors of MMPs, which can prevent MMP activity. The change of flowing bone-marrow-derived cells into cardiac fibroblasts, infiltration of these cells into the heart, and the hyperactivity of these cells, all of which grow in response to particular stressful stimuli, are the primary causes of fibrosis [113]. According to several studies, collagen deposition and fibrosis in cardiac remodeling are associated with elevated levels of collagen synthesis biomarkers (PIIINCP, PIIINP, PICP, and PINP) and lower blood levels of collagen type I degradation biomarker (CITP).

9. Neuroinflammation, neurodegeneration, and cholesterol homeostasis

For cellular homeostasis and transmembrane transfers, both inside and between cellular compartments, cholesterol is crucial. The cholesterol is transferred between neurons in the brain, which have the greatest concentrations of cholesterol in the body, and intracellular organelles to sustain usual brain function. Most of the cholesterol is generated by glial cells; however, neurons can also take up astrocyte-produced cholesterol. Endocytic and retrograde passage routes carry cholesterol to and from the intracellular organelles endosomes and lysosomes. However, it appears that the etiology of Parkinson's disease, Niemann-Pick disease type C illnesses, and aberrant cholesterol trafficking are all related to neuroinflammation, neurodegeneration, and cholesterol homeostasis [114]. Abnormal molecular interactions or specific cholesterol depositions have been seen as important factors that hasten the death of neuronal cells under the diseased settings of these neurodegenerative illnesses [115]. The authors address current developments about the function of cholesterol in a fit brain and the molecular pathways by which Niemann-Pick disease, Alzheimer's, and Parkinson's disorders are impacted by cholesterol. With a focus on cholesterol interactions with Niemann-Pick disease and synuclein in Alzheimer's disorders, Niemann-Pick disease, and Parkinson's disease, authors explore several lines of proof assistant various hypotheses of aberrant intracellular cholesterol operating.

Recently, it was found that brain cells are unable to maintain their protective myelin sheaths in the absence of a protein known as transactive response DNA-binding protein [116]. Studies have demonstrates that the protein transactive response DNA-binding protein, which plays a role in diseases such as frontal lobe dementia and amyotrophic lateral sclerosis, affects cholesterol metabolism in the brain [117]. Moreover, it has been demonstrated how crucial cholesterol synthesis and absorption are for developing myelin sheaths. Researchers have specified that the experimental data they obtained

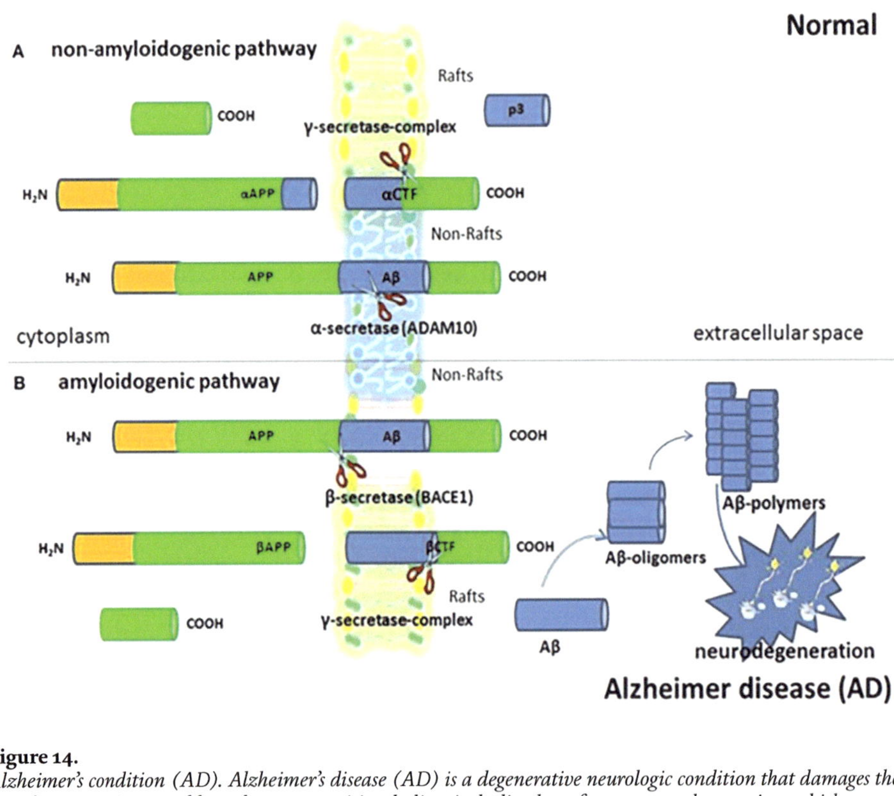

Figure 14.
Alzheimer's condition (AD). Alzheimer's disease (AD) is a degenerative neurologic condition that damages the brain's neurons irreparably and causes cognitive decline, including loss of memory and reasoning, which can make it difficult to operate in social or professional contexts [119]. Reprinted (adapted) with permission from. Copyright (2013) Antonella Mandas et al. Front Physiol, Frontiers.

from their previous work in mice with transactive response DNA-binding protein removed from oligodendrocytes were the basis for their investigation into the relationship between transactive response DNA-binding protein and cholesterol metabolism [118]. Oligodendrocytes shield and myelinate neurons, increasing the transmission speed as a result. The investigators revealed that animals lacking transactive response DNA-binding protein in their oligodendrocytes exhibit increasing neurological abnormalities that result in early mortality (**Figure 14**). It was described that these characteristics were followed by oligodendrocyte death and progressive myelin atrophy.

Chronic neuroinflammation and oxidative stress are two interwoven pathologic variables that contribute to neurodegenerative disorders (such as Parkinson's and Alzheimer's disease) and brain aging. Numerous biological processes depend on reactive oxygen species (ROS), which serve as physiological signaling molecules. However, the redox imbalance that results when the concentration of ROS overpowers the antioxidant defense mechanism compromises cellular reliability and functioning. The brain reacts to oxidative stress rapidly. New treatment approaches to arrest the main causes of oxidative stress in neurodegenerative illnesses are required in light

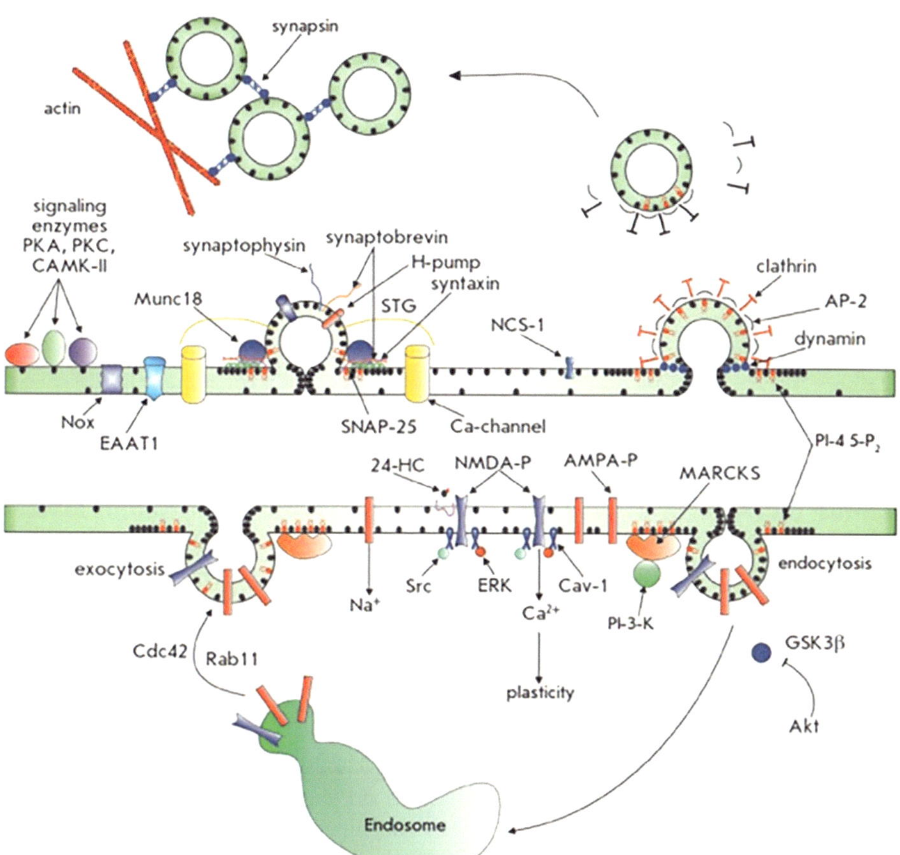

Figure 15.
Lipid-protein interactions in synaptic transmission. When synaptic vesicles fuse (exocytose) with the presynaptic membrane at a particular location (referred to as the active zone) in response to Ca^{2+} inflow via potential-gated Ca^{2+} channels, the neurotransmitter is released. Reprinted (adapted) with permission from [34]. Copyright (2016) A. M. Petrov et al. Acta Naturae, Europe PMC.

of the failure of free-radical-scavenging antioxidants in clinical trials. Although the mitochondrial electron transport chain is the main source of intracellular ROS, chronic neurotoxicity caused by overactive phagocytic NADPH oxidase (NOX2), a significant inflammatory oxidative enzyme, has been shown to be primarily mediated by NOX2 in models of neurodegenerative disorders [120]. Furthermore, recent research has indicated that dysregulated chronic neuroinflammation may be the catalyst and trigger for long-term neurodegenerative processes.

Treatment for neurodegenerative illnesses may be more effective when the confluence of oxidative stress and neuroinflammation is blocked. As a result, the targeting microglial NOX2 may develop into a promising therapeutic approach that alters the course of neurodegenerative disorders. When homeostasis in the brain is disturbed, either directly through the production of aberrant proteins or cerebral hypoperfusion, or indirectly through peripheral inflammation, microglia are stimulated to yield a range of pro-inflammatory constituents that may cause swelling and cell death [121]. Pro-inflammatory cytokines may cause alterations in the iron proteins necessary to maintain iron homeostasis, resulting in increased iron deposition in brain cells. Through their interactions with iron-regulatory proteins, reactive nitrogen and oxygen species, which are intimately associated with the inflammatory process, can have a major impact on iron metabolism [122]. This explains why it is crucial to maintain iron levels in a state that can be controlled; as a result, the sophisticated systems that control iron homeostasis can be better understood (**Figure 15**). N-acetyl cysteine, non-steroidal anti-inflammatory drugs, and iron chelation may be utilized as therapeutic methods to reduce the toxicity of iron in biological systems.

10. Cholesterol homeostasis, and targeted therapeutics

Cholesterol governs cellular homeostasis and maintains the firmness of cell sheaths. Thus, preserving cholesterol homeostasis is crucial for regular cellular activity. By controlling cholesterol production, efflux, and intake from lipoprotein carriers, cellular cholesterol maintains homeostasis. Obesity, diabetes, cancer, and cardiovascular disease are only a few disorders that are typically brought on by abnormal trafficking and cellular cholesterol homeostasis [123]. According to recent studies, several inherited illnesses may also be triggered by a disturbed cholesterol homeostasis. An accumulative quantity of research has shown a direct connection between acquired diseases such as cardiovascular problems, liver problems, and many types of cancer and cholesterol homeostasis. Due to its therapeutic potential as a target in both the prevention and therapy of cancer, cholesterol is receiving more and more interest in cancer research. Its involvement in tumorigenicity, however, is still debatable. Cholesterol homeostasis has been documented as a new essential reason in the pathogenesis of cancer based on the available data.

The role of cholesterol in signal transduction and conversion into other crucial macromolecules such as sterol hormones and bile acids has been underlined. According to research, cholesterol has a distinctly paradoxical function in the development of cancer, demonstrating that the relationship between cholesterol and carcinogenicity might vary depending on the kind of cancer. High cholesterol or hypercholesteremia is associated with breast and prostate cancers but still requires more authentic proof [124]. The key findings of recent pre-clinical and clinical studies are being explored in cholesterol metabolism in cancer. Recent studies have explained

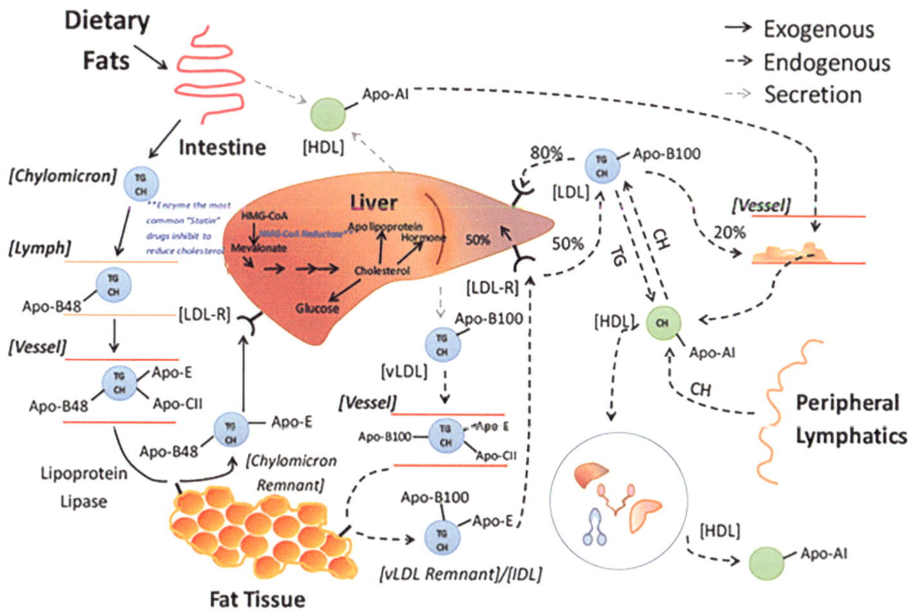

Figure 16.
Lipid metabolism and new treatments to lower cholesterol and prevent cardiovascular disease. Reprinted (adapted) with permission from [125]. Copyright (2016) Monte S. Willis et al. Journal of Cardiology and Therapy, Europe PMC.

the therapeutic role of natural compounds that are applied as cholesterol-lowering medications during the treatment of cancer (**Figure 16**). Worldwide cancer incidence and cholesterol content have found to be linked in epidemiological studies. Evidence is mounting that the disruption of cholesterol metabolism plays a growing role in the emergence of cancer [126]. More precisely, recent studies have demonstrated how cholesterol plays a unique function in the control of cell survival, the inhibition of immune cells, and the variation of cancer stem cells. In conclusion, altering cholesterol metabolism may represent a fresh therapeutic approach for treating cancer.

Interrupted cholesterol homeostasis is a starting point for the onset of many diseases, including cancer, neurodegenerative diseases, and cardiovascular diseases, during these unwanted cellular events, lipid buildup (primarily cholesteryl esters) in macrophage/foam cells beneath the endothelial region eventually leads to atherosclerotic lesions [127]. Various studies have demonstrated that reducing cholesterol levels, particularly those of low-density lipoprotein cholesterol effectively defends cardiovascular system and decreases the threat of cardiovascular happenings [128]. Cholesterol uptake, storage, efflux, use, biosynthesis excretion, and transportation play a role in maintaining cholesterol homeostasis. Many regulatory pathways should precisely govern every process. Numerous therapies have been devised to reduce levels of cholesterol by preventing cholesterol production and uptake or increasing cholesterol use and excretion. These strategies are based on the management of cholesterol homeostasis [129]. **Figure 17** depicts the current knowledge of the molecular pathways important for controlling cholesterol homeostasis, nanotherapies used to lower cholesterol based on clinical or preclinical investigations, and emerging cholesterol-lowering targets.

Reported data point to a clear connection between inflammation, hematological malignancies, and cholesterol homeostasis in addition to maintaining cell integrity

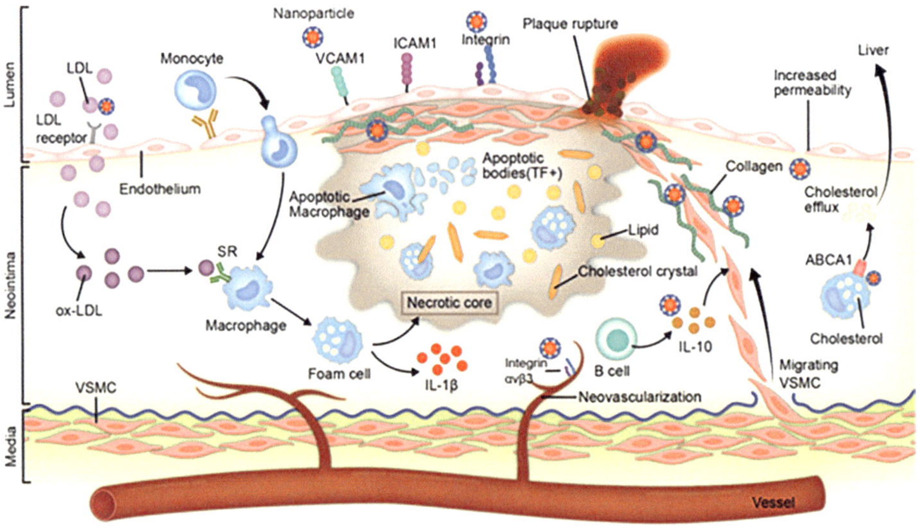

Figure 17.
Nanomedicine-based strategies for targeting advanced atherosclerotic plaques [28]. Reprinted (adapted) with permission from. Copyright (2021) Acta Pharmacol Sin. et al. Acta Pharmacol Sin., Springer nature.

and permeability. As a result, the ideal therapeutic focus for hematological malignancies is cholesterol homeostasis. The integrity of tumor cells is impacted in vitro and in vivo by altering the homeostasis of cholesterol, either by preventing its synthesis or by triggering reverse cholesterol transport by activating liver X receptors [130]. In preclinical animals, cholesterol homeostasis has been altered to revive anticancer immune responses. These findings have led to the testing of chemotherapy and medicines that prevent the synthesis of cholesterol in clinical studies involving acute myeloid leukemia (statins). Body will not be in deprivation of cholesterol, which is crucial for its normal operation, using this method. To assess the downstream implications of the cholesterol-depleting method, which include the effects on various signaling pathways, adverse effects, and long-term treatment, more research is still needed. Clinical trials and *in vivo* research must evaluate the efficacy and validity of this therapeutic approach [18]. Creating high-throughput siRNA displays for the genes implicated in cholesterol signaling routes and paths to determine their connection with innumerable cholesterol-dependent malignancies could be a promising screening technique [131]. It would also be necessary to research how well-known medications affect the levels of gene expression in pathways associated with cholesterol. Over the course of the chapter, the authors examine the possible applications of lipid modulators as therapeutic agents and evaluate the importance of cholesterol homeostasis in hematological distortions as well as in tumor cells and associated microenvironment.

This chapter examines the linkages between cholesterol and the signaling pathways that drive cancer and has underlined the urgent need to lower cholesterol levels in cancer cells to stop unchecked-cell growth that can prevent the progression of treatment battle in cancer cells. The authors propose that lowering cholesterol levels must be investigated as a potential therapeutic target for curing fatal diseases in the ongoing search for alternate therapies [132]. Several ambiguities persist about statins, it can be underlined here clearly that they can be used in clinical trials counter to cancer soon [133]. To further diminish cancer medication resistance, combining a cholesterol-depleting agent with anticancer therapy medicines may be relevant. Generally, this somewhat new domain of exploration presents exciting prospects to the researchers for new anticancer drug development as well as understanding the biology of cholesterol, particularly concerning cancer at the system level, to develop new strategies to combat cancer drug resistance. Thus, the emerging potentials of cholesterol regulation and its implications in the management of various disease conditions must be explored.

Acknowledgements

For inspiration, the author (Rajiv Kumar) heartily acknowledges his younger brother Bitto.

Author contributions

Dr. Rajiv Kumar drafted the work and figures. Dr. Neelam Chhillar, Dr. Shailey Singhal, and Dr. Tanya Chauhan, Dr. Ginpreet Kaur, and Dhruv Sanjay Gupta revised it critically for important intellectual content and approved the final version.

Competing interests

The author has stated that they have no conflicting interests.

Availability of data and materials

Where applicable, the reference section contains pertinent citations.

Author details

Rajiv Kumar[1,2]*, Neelam Chhillar[3], Dhruv Sanjay Gupta[4], Ginpreet Kaur[4], Shailey Singhal[5], Tanya Chauhan[6] and Sandeep Mittan[7]

1 University of Delhi, Delhi, India

2 NIET, National Institute of Medical Science, India

3 Deparetment of Biochemistry, School of Medicine, DY Patil University, Navi Mumbai, India

4 Department of Pharmacology, SPP School of Pharmacy and Technology Management, SVKM's NMIMS, Mumbai, India

5 Cluster of Applied Sciences, University of Petroleum and Energy Studies, Dehradun, India

6 Division of Forensic Biology, National Forensic Sciences University, Delhi, India

7 Division of Cardiovascular Research, Albert Einstein College of Medicine, New York, NY, USA

*Address all correspondence to: chemistry_rajiv@hotmail.com

References

[1] Linton MF et al. The role of lipids and lipoproteins in atherosclerosis. Science (80-.). 2019;**111**:80-85

[2] Kumar R, Gulia K. The convergence of nanotechnology-stem cell, Nano topography-mechanobiology, and biotic-abiotic interfaces: Nanoscale tools for tackling the top killer, arteriosclerosis, strokes, and heart attacks. Nano Select. 2021;**2**:655-687

[3] Wunderer F et al. The role of hepcidin and iron homeostasis in atherosclerosis. Pharmacological Research. 2020;**153**:104664

[4] Rajiv K. Physiology, coagulation Cascade: Inherited disorders, and the molecular phenomenon of alterations in hemostasis. Journal of Clinical Haematology. 2021;**2**:62-64

[5] Rajiv K. Biomedical applications of nanoscale tools and nano-bio interface: A blueprint of physical, chemical, and biochemical cues of cell mechanotransduction machinery. Biomedical Research and Clinical Reviews. 2021;**4**(64):1-4

[6] Yudin D, Fainzilber M. Ran on tracks—Cytoplasmic roles for a nuclear regulator. Journal of Cell Science. 2009;**122**:587-593

[7] Gulia RK. Cell mechanotransduction machinery, and cell signaling defects: Small tools and nano-bio interface for influential regenerative remedies. Journal of Cellular Signaling. 2021;**6**:1-14

[8] Singh P, Saxena R, Srinivas G, Pande G, Chattopadhyay A. Cholesterol biosynthesis and homeostasis in regulation of the cell cycle. PLoS One. 2013;**8**:e58833

[9] Cerqueira NMFSA et al. Cholesterol biosynthesis: A mechanistic overview. Biochemistry. 2016;**55**:5483-5506

[10] Kumar N, Kumar R. Nanotechnology and Nanomaterials in the Treatment of Life-Threatening Diseases. Elsevier; 2013. DOI: 10.1016/C2013-0-13374-0

[11] Go GW, Mani A. Low-density lipoprotein receptor (LDLR) family orchestrates cholesterol homeostasis. Yale Journal of Biology and Medicine. 2012;**85**:19-28

[12] Duan Y et al. Regulation of cholesterol homeostasis in health and diseases: From mechanisms to targeted therapeutics. Signal Transduction and Targeted Therapy. 2022;**7**:265

[13] Rajiv K. Traumatic brain injury: Mechanistic insight on pathophysiological mechanisms underlying, neurotransmitters, and potential therapeutic targets. Medical & Clinical Reviews. 2021;**7**:1-3

[14] Phillips MC, Johnson WJ, Rothblat GH. Mechanisms and consequences of cellular cholesterol exchange and transfer. BBA—Reviews on Biomembranes. 1987;**906**:223-276

[15] Schirone L et al. A review of the molecular mechanisms underlying the development and progression of cardiac remodeling. Oxidative Medicine and Cellular Longevity. 2017;**2017**:3920195

[16] Malekmohammad K, Bezsonov EE, Rafieian-Kopaei M. Role of lipid accumulation and inflammation in atherosclerosis: Focus on molecular and cellular mechanisms. Frontiers in Cardiovascular Medicine. 2021;**8**:707529

[17] Rajiv K. Elucidation of the origin of autoimmune diseases via computational multiscale mechanobiology and extracellular matrix remodeling: Theories

and phenomenon of immunodominance. Current Medical and Drug Research. 2021;**5**:215

[18] Kumar R, Sandeep Mittan JS. Nanobiomaterials, nanobiomechanics and tissue bioengineering for advanced regenerative therapeutics: Present and future perspectives. Journal of Materials Nanoscience. 2015;**2**:15-26

[19] Pietrocola F, Galluzzi L, Bravo-San Pedro JM, Madeo F, Kroemer G. Acetyl coenzyme A: A central metabolite and second messenger. Cell Metabolism. 2015;**21**:805-821

[20] Rajiv K. Cell shrinkage, cytoskeletal pathologies, and neurodegeneration: Myelin sheath formation and remodeling. Archives of Medical and Clinical Research. 2021;**1**:1-5

[21] Prinsen BHCMT et al. Endogenous cholesterol synthesis is associated with VLDL-2 apoB-100 production in healthy humans. Journal of Lipid Research. 2003;**44**:P1341-1348

[22] Mani V, Park S, Kim JA, Lee SI, Lee K. Metabolic perturbation and synthetic biology strategies for plant terpenoid production—An updated overview. Plants. 2021;**10**:2179

[23] Pinatel E et al. Interplay between non-coding rna transcription, stringent/relaxed phenotype and antibiotic production in streptomyces ambofaciens. Antibiotics. 2021;**10**:947

[24] Rather MA et al. Inflammation and Alzheimer's disease: Mechanisms and therapeutic implications by natural products. Mediators of Inflammation. 2021;**2021**:9982954

[25] Daniels TF, Killinger KM, Michal JJ, Wright RW, Jiang Z. Lipoproteins, cholesterol homeostasis and cardiac health. International Journal of Biological Sciences. 2009;**5**:474-488

[26] Silva Afonso M et al. Molecular pathways underlying cholesterol homeostasis. Nutrients. 2018;**10**:760

[27] Saeed A et al. Remnant-like particle cholesterol, low-density lipoprotein triglycerides, and incident cardiovascular disease. Journal of the American College of Cardiology. 2018;**72**:156-169

[28] Ou LC, Zhong S, Ou JS, Tian JW. Application of targeted therapy strategies with nanomedicine delivery for atherosclerosis. Acta Pharmacologica Sinica. 2021;**42**:10-17

[29] Matthan NR et al. Cholesterol absorption and synthesis markers in individuals with and without a CHD event during pravastatin therapy: Insights from the PROSPER trial. Journal of Lipid Research. 2010;**51**:202-209

[30] Lu K, Lee MH, Patel SB. Dietary cholesterol absorption; more than just bile. Trends in Endocrinology and Metabolism. 2001;**12**:P314-320

[31] Bowman MP, Van Doren J, Taper LJ, Thye FW, Ritchey SJ. Effect of dietary fat and cholesterol on plasma lipids and lipoprotein fractions in normolipidemic men. The Journal of Nutrition. 1988;**118**:555-560

[32] Jones PJH et al. Dietary cholesterol feeding suppresses human cholesterol synthesis measured by deuterium incorporation and urinary mevalonic acid levels. Arteriosclerosis, Thrombosis, and Vascular Biology. 1996;**16**:1222-1228

[33] Chen HC. Molecular mechanisms of sterol absorption. The Journal of Nutrition. 2001;**131**:2603-2605

[34] Petrov AM, Kasimov MR, Zefirov AL. Brain cholesterol metabolism and its defects: Linkage to neurodegenerative diseases and synaptic dysfunction. Acta Naturae. 2016;**8**:239-259

[35] Davis HR, Altmann SW. Niemann-Pick C1 Like 1 (NPC1L1) an intestinal sterol transporter. Biochimica et Biophysica Acta—Molecular and Cell Biology of Lipids. 2009;**1791**:58-73

[36] Yu XH et al. ABCG5/ABCG8 in cholesterol excretion and atherosclerosis. Clinica Chimica Acta. 2014;**428**:82-88

[37] Kumar N, Wang H, Liu D, Collins S. Liver X receptor is a regulator of orphan nuclear receptor NOR-1 gene transcription in adipocytes. International Journal of Obesity. 2009;**33**:519-524

[38] Zhao C, Dahlman-Wright K. Liver X receptor in cholesterol metabolism. Journal of Endocrinology. 2010;**204**:233-240

[39] Bilotta MT, Petillo S, Santoni A, Cippitelli M. Liver X receptors: Regulators of cholesterol metabolism, inflammation, autoimmunity, and cancer. Frontiers in Immunology. 2020;**11**:584303

[40] American Heart Association. Prevention and treatment of high cholesterol (hyperlipidemia). American Heart Association. 2017;**501**. Web page

[41] Ahmed MH. Ezetimibe and recent clinical trials: A look on the bright side. Expert Opinion on Drug Safety. 2010;**9**:511-514

[42] Shi Q, Chen J, Zou X, Tang X. Intracellular cholesterol synthesis and transport. Frontiers in Cell and Development Biology. 2022;**10**:Art ID 819281

[43] Temel RE, Mark Brown J. A new framework for reverse cholesterol transport: Non-biliary contributions to reverse cholesterol transport. World Journal of Gastroenterology. 2010;**16**:5946-5952

[44] Soffientini U, Graham A. Intracellular cholesterol transport proteins: Roles in health and disease. Clinical Science. 2016;**130**:1843-1859

[45] Tisdale MJ. Role of acetoacetyl-CoA synthetase in acetoacetate utilization by tumor cells. Cancer Biochemistry Biophysics. 1984;7:101-107

[46] Cholesterol: Synthesis, Metabolism, and Regulation. themedicalbio chemistrypage, LLC; 2022. Available from: https://themedicalbiochemistrypage.org/cholesterol-synthesis-metabolism-and-regulation/

[47] Wu N et al. Activation of 3-hydroxy-3-methylglutaryl coenzyme A (HMG-CoA) reductase during high fat diet feeding. Biochimica et Biophysica Acta, Molecular Basis of Disease. 2013;**1832**:1560-1568

[48] Theesfeld CL, Pourmand D, Davis T, Garza RM, Hampton RY. The sterol-sensing domain (SSD) directly mediates signal-regulated endoplasmic reticulum-associated degradation (ERAD) of 3-hydroxy-3- methylglutaryl (HMG)-CoA reductase isozyme Hmg2. The Journal of Biological Chemistry. 2011;**286**:P26298-26307

[49] Dong XY, Tang SQ. Insulin-induced gene: A new regulator in lipid metabolism. Peptides. 2010;**31**:2145-2150

[50] Guo F et al. Upregulation of 24(R/S),25-epoxycholesterol and 27-hydroxycholesterol suppresses the proliferation and migration of gastric cancer cells. Biochemical and Biophysical Research Communications. 2018;**504**:892-898

[51] Radhakrishnan A, Ikeda Y, Hyock JK, Brown MS, Goldstein JL. Sterol-regulated transport of SREBPs from endoplasmic reticulum to Golgi: Oxysterols block transport by binding to Insig. Proceedings of the National Academy of Sciences of the United States of America. 2007;**104**:6519-6526

[52] Kostner GM et al. HMG CoA reductase inhibitors lower LDL cholesterol without reducing Lp(a) levels. Circulation. 1989;**80**:497-510

[53] Civra A et al. 25-Hydroxycholesterol and 27-hydroxycholesterol inhibit human rotavirus infection by sequestering viral particles into late endosomes. Redox Biology. 2018;**19**:318-330

[54] Ni L, Yuan C. The mitochondrial-associated endoplasmic reticulum membrane and its role in diabetic nephropathy. Oxidative Medicine and Cellular Longevity. 2021;**2021**:8054817

[55] Townley AK et al. Efficient coupling of Sec23-Sec24 to Sec13-Sec31 drives COPII-dependent collagen secretion and is essential for normal craniofacial development. Journal of Cell Science. 2008;**121**:3025-3034

[56] Danyukova T, Schöneck K, Pohl S. Site-1 and site-2 proteases: A team of two in regulated proteolysis. Biochimica et Biophysica Acta, Molecular Cell Research. 2022;**1869**:119138

[57] Criado-Mesas L et al. Expression analysis of porcine miR-33a/b in liver, adipose tissue and muscle and its potential role in fatty acid metabolism. PLoS One. 2021;**16**:e0245858

[58] Steck TL, Lange Y. SCAP, an ER sensor that regulates cell cholesterol. Developmental Cell. 2002;**3**:724-729

[59] Song BL, Sever N, DeBose-Boyd RA. Gp78, a membrane-anchored ubiquitin ligase, associates with Insig-1 and couples sterol-regulated ubiquitination to degradation of HMG CoA reductase. Molecular Cell. 2005;**19**:829-840

[60] Rawson RB. The SREBP pathway—Insights from insigs and insects. Nature Reviews Molecular Cell Biology. 2003;**4**:631-640

[61] Kaneda Y et al. Antimetastatic effect of synthetic Glu-Ile-Leu-Asp-Val peptide derivatives containing D-amino acids. Anti-Cancer Drugs. 1997;**8**:702-707

[62] Cook ECL et al. Identification of the ER-resident E3 ubiquitin ligase RNF145 as a novel LXR-regulated gene. PLoS One. 2017;**12**:e0172721

[63] Cholesterol: Synthesis, Metabolism, and Regulation. themedicalbiochemistrypage, LLC. Available from: https://themedicalbiochemistrypage.org/cholesterol-synthesis-metabolism-and-regulation/

[64] Chua NK, Coates HW, Brown AJ. Squalene monooxygenase: A journey to the heart of cholesterol synthesis. Progress in Lipid Research. 2020;**79**:1-27

[65] Lee YC et al. Znf179 E3 ligase-mediated TDP-43 polyubiquitination is involved in TDP-43- ubiquitinated inclusions (UBI) (+)-related neurodegenerative pathology 11 Medical and Health Sciences. Journal of Biomedical Science. 2018;**25**. Webpage

[66] Taylor WC, Pass TM, Shepard DS, Komaroff AL. Cholesterol reduction and life expectancy. A model incorporating multiple risk factors. Annals of Internal Medicine. 1987;**106**:605-614

[67] Willey JZ, Elkind MSV. 3-Hydroxy-3-methylglutaryl-coenzyme A reductase inhibitors in the treatment of central nervous system diseases. Archives of Neurology. 2010;**67**:1062-1067

[68] Nestel PJ, Poyser A. Changes in cholesterol synthesis and excretion when cholesterol intake is increased. Metabolism. 1976;**25**:P1591-1599

[69] Vessey DA. The biochemical basis for the conjugation of bile acids with either glycine or taurine. The Biochemical Journal. 1978;**174**:621-626

[70] Chiang JYL. Negative feedback regulation of bile acid metabolism: Impact on liver metabolism and diseases. Hepatology. 2015;**62**:1315-1317

[71] Friesen JA, Rodwell VW. The 3-hydroxy-3-methylglutaryl coenzyme A (HMG-CoA) reductases. Genome Biology. 2004;**5**:248

[72] Reimão S et al. 3-Hydroxy-3-methylglutaryl-coenzyme A lyase deficiency: Initial presentation in a young adult. Journal of Inherited Metabolic Disease. 2009;**32**:49-52

[73] Slezak-Prochazka I, Selvi D, Kroesen BJ, Van Den Berg A. MicroRNAs, macrocontrol: Regulation of miRNA processing. RNA. 2010;**16**:1087-1095

[74] Dimarco DM, Fernandez ML. The regulation of reverse cholesterol transport and cellular cholesterol homeostasis by microRNAs. Biology. 2015;**4**:494-511

[75] Gelissen IC, Brown AJ. An overview of cholesterol homeostasis. Methods in Molecular Biology. 2017;**1583**:1-6

[76] Marcel YL, Ouimet M, Wang MD. Regulation of cholesterol efflux from macrophages. Current Opinion in Lipidology. 2008;**19**:455-61

[77] Rothblat GH, Phillips MC. Cholesterol efflux: Mechanism and regulation. Advances in Experimental Medicine and Biology. 1986;**201**:195-204

[78] Chen W, Silver DL, Smith JD, Tall AR. Scavenger receptor-BI inhibits ATP-binding cassette transporter 1-mediated cholesterol efflux in macrophages. The Journal of Biological Chemistry. 2000;**275**:P30794-30800

[79] Chauhan T, Rajiv K. Molecular markers and their applications in fisheries and aquaculture. Advances in Bioscience and Biotechnology. 2010;**01**:281-291

[80] Juhl AD, Wüstner D. Pathways and mechanisms of cellular Cholesterol efflux—Insight from imaging. Frontiers in Cell and Development Biology. 2022;**10**:359-364

[81] Liem AH et al. ApoB/A1 and LDL-C/HDL-C and the prediction of cardiovascular risk in statin-treated patients. Current Medical Research and Opinion. 2008;**24**:536-542

[82] Siddiqi HK, Kiss D, Rader D. HDL-cholesterol and cardiovascular disease: Rethinking our approach. Current Opinion in Cardiology. 2015;**30**:834408

[83] Ouimet M, Wang MD, Cadotte N, Ho K, Marcel YL. Epoxycholesterol impairs cholesteryl ester hydrolysis in macrophage foam cells, resulting in decreased cholesterol efflux. Arteriosclerosis, Thrombosis, and Vascular Biology. 2008;**28**:1144-1150

[84] Rajiv K. A mechanistic insight on pathophysiological mechanisms of inflammatory diseases and potential therapeutic targets. Journal of Scientific Research and Biomedical Informatics. 2022;**3**:1-4

[85] Sakai J, Rawson RB. The sterol regulatory element-binding protein pathway: Control of lipid homeostasis through regulated intracellular transport. Current Opinion in Lipidology. 2001;**12**:261-266

[86] Krasowski MD, Ni A, Hagey LR, Ekins S. Evolution of promiscuous nuclear hormone receptors: LXR, FXR, VDR, PXR, and CAR. Molecular and Cellular Endocrinology. 2011;**334**:39-48

[87] Li G, Gu HM, Zhang DW. ATP-binding cassette transporters and cholesterol translocation. IUBMB Life. 2013;**65**:505-512

[88] BastanR,EskandariN,SabzghabaeeAM, Manian M. Serine/threonine phosphatases:

Classification, roles and pharmacological regulation. International Journal of Immunopathology and Pharmacology. 2014;**27**:473-484

[89] Anastasius M et al. Cholesterol efflux capacity: An introduction for clinicians. American Heart Journal. 2016;**180**:54-63

[90] Kondo A et al. Extracellular acidic pH activates the sterol regulatory element-binding protein 2 to promote tumor progression. Cell Reports. 2017;**18**: 2228-2242

[91] Dorotea D, Koya D, Ha H. Recent insights into SREBP as a direct mediator of kidney fibrosis via lipid-independent pathways. Frontiers in Pharmacology. 2020;**11**:265

[92] Horton JD, Goldstein JL, Brown MS. SREBPs: Activators of the complete program of cholesterol and fatty acid synthesis in the liver. Journal of Clinical Investigation. 2002;**109**:1125-1131

[93] Luu W, Sharpe LJ, Gelissen IC, Brown AJ. The role of signalling in cellular cholesterol homeostasis. IUBMB Life. 2013;**65**:159-191

[94] Kalaany NY, Mangelsdorf DJ. LXRs and FXR: The Yin and Yang of cholesterol and fat metabolism. Annual Review of Physiology. 2006;**68**:675-684

[95] Ignatova ID, Angdisen J, Moran E, Schulman IG. Differential regulation of gene expression by LXRs in response to macrophage cholesterol loading. Molecular Endocrinology. 2013;**27**:1036-1047

[96] Patel SB, Graf GA, Temel RE. ABCG5 and ABCG8: More than a defense against xenosterols. Journal of Lipid Research. 2018;**59**:P1103-1113

[97] Bettzieche A, Brandsch C, Hirche F, Eder K, Stangl GI. L-cysteine down-regulates SREBP-1c-regulated lipogenic enzymes expression via glutathione in HepG2 cells. Annals of Nutrition & Metabolism. 2008;**52**:196-203

[98] Katz SS, Small DM, Smith FR, Dell RB, Goodman DS. Cholesterol turnover in lipid phases of human atherosclerotic plaque. Journal of Lipid Research. 1982;**23**:733-737

[99] Kutsky P, Parker JL. Calcium fluxes in cardiac sarcolemma and sarcoplasmic reticulum isolated from endotoxin-shocked Guinea pigs. Circulatory Shock. 1990;**30**:349-64

[100] Tang J, Xie Q, Ma D, Wang W. Effects of ET-1 and TNF-α levels on the cardiac function and prognosis in rats with chronic heart failure. European Review for Medical and Pharmacological Sciences. 2019;**23**:11004-11010

[101] Rajiv K. Brain ischemia, and reperfusion injury: A clinical and physiological investigation of neuroinflammation and chronic neuroinflammatory diseases. British Journal of Pharmaceutical and Medical Research. 2022;7:3539-3543

[102] Zelová H, Hošek J. TNF-α signalling and inflammation: Interactions between old acquaintances. Inflammation Research. 2013;**62**:641-651

[103] Kumar R, Chhillar M. Nano-bio interface, bioadaptability of different nanoparticles, nanokicking and extracellular matrix mimicking: A biological and medicinal front to promote the concept of a cell, having a better defense system inbuilt by nature. Aditum Journal of Clinical and Biomedical Research. 2021;**1**:1-7

[104] Randall G, Molloy GJ, Steptoe A. The impact of an acute cardiac event on the partners of patients: A systematic review. Health Psychology Review. 2009;**3**:1-84

[105] Kumar R. Ischemia-reperfusion injury: A mechanistic concept. Atherosclerosis Open Access. 2021;**6**:1-2

[106] Fedak PWM, Verma S, Weisel RD, Li RK. Cardiac remodeling and failure: From molecules to man (part I). Cardiovascular Pathology. 2005;**14**:1-11

[107] Kumar R. Emerging role of neutrophils in wound healing and tissue repair: The routes of healing. Biomedical Journal of Scientific & Technical Research. 2021;**36**:28687-28688

[108] D., H. et al. Renal sympathetic nerve ablation in moderate to severe chronic renal failure: A short term safety and efficacy pilot study. Hypertension. 2012;**60**:39-46

[109] Carter J, Lippa C. β-Amyloid, neuronal death and Alzheimers disease. Current Molecular Medicine. 2005;**1**:733-737

[110] Alexandrou ME et al. Cardiac geometry, function, and remodeling patterns in patients under maintenance hemodialysis and peritoneal dialysis treatment. Therapeutic Apheresis and Dialysis. 2022;**26**:601-612

[111] Kumar R, Chhikara BS, Gulia K, Chhillar M. Cleaning the molecular machinery of cellsviaproteostasis, proteolysis and endocytosis selectively, effectively, and precisely: Intracellular self-defense and cellular perturbations. Molecular Omics. 2021;**17**:11-28

[112] Candelario-Jalil E, Yang Y, Rosenberg GA. Diverse roles of matrix metalloproteinases and tissue inhibitors of metalloproteinases in neuroinflammation and cerebral ischemia. Neuroscience. 2009;**158**:983-994

[113] Jang HS, Kim JI, Han SJ, Park KM. Recruitment and subsequent proliferation of bone marrow-derived cells in the postischemic kidney are important to the progression of fibrosis. American Journal of Physiology. Renal Physiology. 2014;**306**:F1451-F1461

[114] Bi X, Liao G. Cholesterol in Niemann–Pick type C disease. Sub-Cellular Biochemistry. 2010;**51**:319-335

[115] Torres S, García-Ruiz CM, Fernandez-Checa JC. Mitochondrial cholesterol in Alzheimer's disease and Niemann–Pick Type C disease. Frontiers in Neurology. 2019;**10**:1168

[116] Rajiv K. Mechanosensors, and mechanosensing: Mechanosensation, a perception of the force and response. Global Journal of Medical Research: (C) Microbiology & Pathology. 2021;**21**:1-3

[117] Gendron TF, Josephs KA, Petrucelli L. Review: Transactive response DNA-binding protein 43 (TDP-43): Mechanisms of neurodegeneration. Neuropathology and Applied Neurobiology. 2010;**36**:97-112

[118] Ory DS. Niemann-Pick type C: A disorder of cellular cholesterol trafficking. Biochimica et Biophysica Acta—Molecular and Cell Biology of Lipids. 2000;**1529**:1-3

[119] Anchisi L, Dessì S, Pani A, Mandas A. Cholesterol homeostasis: A key to prevent or slow down neurodegeneration. Frontiers in Physiology. 2013;**3**:3-15

[120] Zhao RZ, Jiang S, Zhang L, Yu ZB. Mitochondrial electron transport chain, ROS generation and uncoupling (review). International Journal of Molecular Medicine. 2019;**44**:486

[121] Diebold BA, Smith SME, Li Y, Lambeth JD. NOX2 as a target for drug development: Indications, possible complications, and progress.

Antioxidants and Redox Signaling. 2015;**23**:375-405

[122] Salisbury D, Bronas U. Reactive oxygen and nitrogen species: Impact on endothelial dysfunction. Nursing Research. 2015;**64**:53-66

[123] Eyre H et al. Preventing cancer, cardiovascular disease, and diabetes: A common agenda for the American Cancer Society, the American Diabetes Association, and the American Heart Association. Circulation. 2004;**109**:3244-3255

[124] Magura L et al. Hypercholest erolemia and prostate cancer: A hospital-based case-control study. Cancer Causes & Control. 2008;**19**:1259-1266

[125] Yipin Han MSW. The role of PCSK9 in lipid metabolism and its relationship to new therapies for lowering cholesterol and reducing cardiac disease. Journal of Cardiology Therapy. 2015;**2**:393-399

[126] Cruz PMR, Mo H, McConathy WJ, Sabnis N, Lacko AG. The role of cholesterol metabolism and cholesterol transport in carcinogenesis: A review of scientific findings, relevant to future cancer therapeutics. Frontiers in Pharmacology. 2013;**4**:393-399

[127] Rajiv K. Macrophage subtypes, phenotypes, inflammatory molecules, cytokines, and atherosclerotic lesions—Atherosclerosis, metabolic diseases, and pathogenesis, the therapeutic challenges. Journal of Clinical & Experimental Pathology. 2021;**11**:1-3

[128] Kawamoto R, Kikuchi A, Akase T, Ninomiya D, Kumagi T. Low density lipoprotein cholesterol and all-cause mortality rate: Findings from a study on Japanese community-dwelling persons. Lipids in Health and Disease. 2021;**20**:105

[129] Lyu J, Yang EJ, Shim JS. Cholesterol trafficking: An emerging therapeutic target for angiogenesis and cancer. Cells. 2019;**8**:Art ID 389

[130] Brendolan A, Russo V. Targeting cholesterol homeostasis in hematopoietic malignancies. Blood. 2022;**139**:165-176

[131] Thaker NG et al. Designing, optimizing, and implementing high-throughput siRNA genomic screening with glioma cells for the discovery of survival genes and novel drug targets. Journal of Neuroscience Methods. 2010;**185**:204-212

[132] Wu G et al. Overcoming treatment resistance in cancer: Current understanding and tactics. Cancer Letters. 2017;**387**:69-76

[133] Farooqi MAM et al. Statin therapy in the treatment of active cancer: A systematic review and meta-analysis of randomized controlled trials. PLoS One. 2018;**13**:e0209486

Chapter 5

Insights into the Wnt Signaling Pathway Evolution

Elham Rismani, Nasrin Haghighi-Najafabadi, Babak Elyasi Far, Behzad Shahbazi and Ladan Mafakher

Abstract

Animals' Wnt signaling pathways are highly preserved signal transduction pathways, which play a crucial role in embryogenesis and adult tissue homeostasis. This chapter reviews the three major Wnt pathways, focusing on some critical proteins in the Wnt/β-catenin path in terms of their evolution and role in homeostasis. Wnt proteins act as a gateway between extracellular, cytoplasmic, and nuclear components to transmit signaling pathways. The Frizzled (FZD) family as G-protein-coupled receptors activates the signaling pathways by binding to Wnt ligands. LRP5/6, members of the family of low-density lipoprotein receptors (LDLR), associate with FZD receptor and Wnt ligands as co-receptors to initiate the Wnt/β-catenin pathway. The Wnt/β-catenin pathway is regulated by antagonists such as the Dickkopf and secreted Frizzled-related protein (SFRP) families.

Keywords: Wnt signaling pathway, frizzled, Wnt, LRP, Dickkopf, SFRP

1. Introduction

The Wnt signaling pathway activates numerous proteins and transmits biological signals into a cell through cell surface receptors, a constant mechanism in all animal kingdom phyla [1]. In the 1980s, studies on oncogenic retroviruses in colon and breast tumors and identifying the Wingless (Wg) gene in *Drosophila* led to the discovery of Wnt signaling [2]. The Wnt signaling route consists primarily of three signal transduction pathways: the canonical Wnt pathway, the noncanonical planar cell polarity (PCP) pathway, and the noncanonical Wnt/calcium pathway. The categories are based on the presence of the protein beta-catenin (β-catenin) in the canonical pathway, whereas the noncanonical pathway operates independently [3]. The central function of WNT signaling in controlling embryonic progression and adult tissue homeostasis is well understood. From primitive metazoans to humans, WNT signaling pathway elements comprising ligands and receptor genes have duplicated and diversified to establish a robust regulatory response [4]. This chapter comprehensively introduces the Wnt signaling pathway and looks at the critical proteins in terms of their evolution and role in homeostasis.

IntechOpen

1.1 The Wnt/β-catenin signaling pathway

The Wnt/β-catenin pathway comprises four protein groups: the extracellular signal ligands, membrane proteins, cytoplasmic members, and nuclear proteins (**Figure 1**). For example, secreted Frizzled-related protein (SFRP) and Dickkopf (DKK) are the family of regulatory proteins in the Wnt signaling pathway that make up the majority of extracellular signals. The cell membrane proteins include the Wnt receptors Frizzled (specific sevenfold transmembrane receptor Frizzled protein) and lipoprotein receptor-related protein (LRP5/6). The cytoplasmic members consist of β-catenin, the Disheveled family (DVL), glycogen synthase kinase-3β (GSK-3β), adenomatous polyposis coli (APC), axis inhibition protein (AXIN), and casein kinase I (CK1). The translocated β-catenin to the nucleus, T cell-specific transcription factor (TCF)/ lymphoid enhancer-binding factor (LEF)

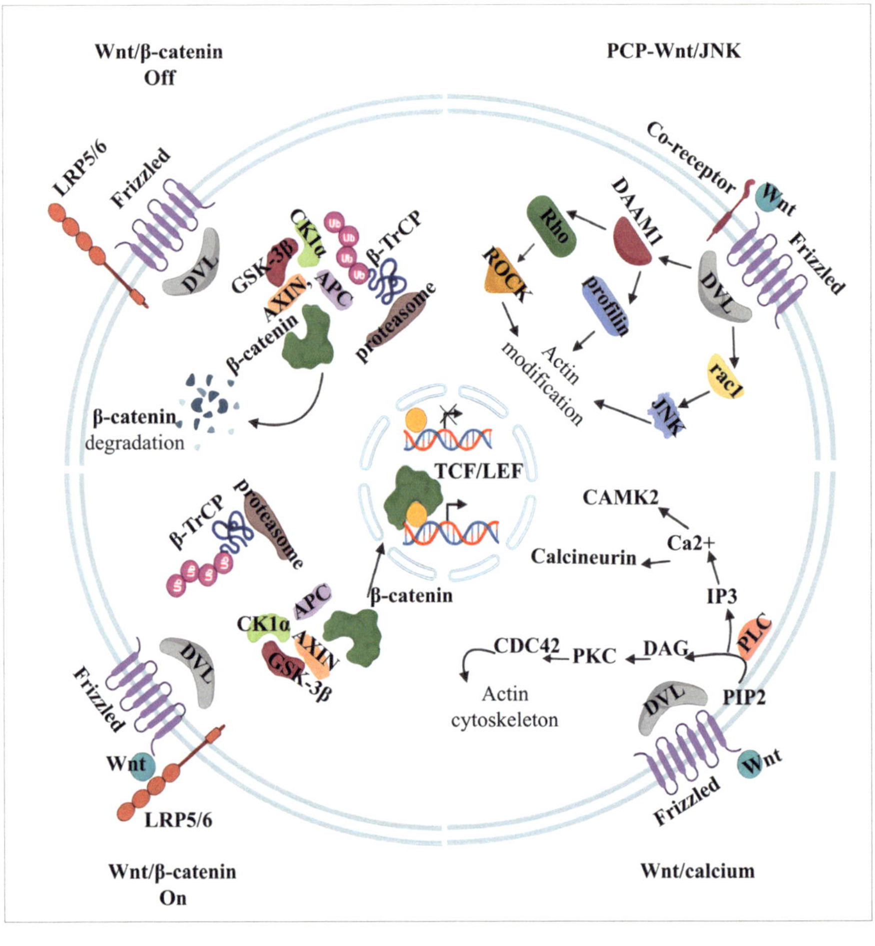

Figure 1.
A schematic representation of the Wnt signaling pathway. In the absence of Wnt ligand (off), β-catenin is targeted for degradation by the Axin, APC, and GSK-3 degradation complex. If Wnt attaches to the Frizzled and LRP receptors (on), the degradation complex is inactivated, causing β-catenin to be stabilized and translocated to the nucleus. The other two pathways function independently of β-catenin. The PCP pathway begins with Wnt molecules interacting with Frizzled and its co-receptor. Attaching Wnt molecules to the Frizzled protein turns on the Wnt/calcium pathway, which can cause calcium to be released from inside the cell.

family members, and downstream target genes, such as MMPs and c-Myc, are members of nuclear proteins [5, 6].

The cytoplasmic level of β-catenin is regulated by the AXIN, APC, CK1, and GSK-3 destruction complex, which causes β-catenin phosphorylation and ubiquitination by beta-transducin repeat-containing E3 ubiquitin-protein ligase (b-TrCP), followed by proteasomal degradation of β-catenin [7, 8]. The binding of Wnt ligands to membrane receptors triggers the binding of AXIN to phosphorylated LRPs, which in turn activates the Wnt/β-catenin pathway. This event makes the destructive complex break apart, stabilizing the cytoplasm's B-catenin and turning on DVL. The primary function of the Wnt/β-catenin pathway comprises the nuclear translocation of the critical protein β-catenin and engagement of target genes via T cell-specific transcription factor and lymphoid enhancer-binding factor (TCF/LEF) [9].

1.2 The planar cell polarity (PCP) signaling pathway

Mutagenesis investigations of Wnt signaling components in Drosophila, particularly Frizzled and Disheveled, revealed that the noncanonical PCP route is fundamental in regulating the posture of epithelial structures such as cuticle hairs and sensory bristles (**Figure 1**). This pathway feature is due to its role in regulating the actin cytoskeleton [10]. The PCP pathway begins with Wnt ligands interacting with Frizzled and its co-receptor (NRH1, Ryk, PTK7, or ROR2). After DVL is recruited, it makes a structure with a Disheveled-associated activator of morphogenesis 1 (DAAM1), which triggers the G-protein Rho and then the Rho-associated kinase. The DVL-Rac1 complex stimulates c-Jun N-terminal kinase (JNK) and facilitates profilin adhesion to actin, which ultimately leads to cytoskeleton remodeling and gastrulation [11, 12].

1.3 The Wnt/calcium signaling pathway

Wnt ligand binding to the Frizzled binding site activates the noncanonical Wnt/calcium cascade (**Figure 1**). The PDZ, DEP domains of DVL, and a G-protein interact directly with Frizzled receptors, stimulating either PLC or cGMP-specific PDE. Activating PLC cleavage of PIP2 as the plasma membrane component into DAG and IP3 triggers calcium release by binding IP3 to its receptor on the ER. Calcium and DAG, along with PKC, can activate CDC42. Calcineurin and CaMKII are also activated by calcium; the latter activates the transcription factor NFAT, which regulates cell attachment, migration, and tissue dissociation. Calcium release is inhibited by activation of PDE and mediates the inhibition of PKG [12, 13].

2. The Wnt signaling pathway's function and role

In the world of multicellular creatures, all three pathways have multiple functions. One of the crucial functions of Wnt signaling is in the orientation of the body axis of the developing embryos in all metazoans [14]. Wnt signaling is critical in embryonic development and controls processes like cell fate decisions, body orientation architecture, cell growth, cell migration, cellular differentiation, and tissue patterning. Wnt signaling appears to regulate the differentiation of embryonic stem cells (ESCs) into different germ layers, consistent with its role in embryonic development. Wnt signaling is an essential regulatory pathway during adult tissue homeostasis in multicellular animals. The involvement of the components of this pathway has been recognized in homeostatic tissue regeneration in the developing and adult brain, metabolic homeostasis, skin homeostasis, and bone homeostasis. Examples come from various studies summarized in the following, considering their references to indicate more details.

One of the most studied cases of the Wnt pathway in the homeostasis of mature tissue is related to the preservation of ESCs niches, whose proliferation and self-renewal are driven by exogenous Wnt signals. Disrupting some components, such as Tcf4, β-catenin, or APC, results in a loss of intestinal epithelial compartment proliferation and differentiation, confirming the importance of Wnt/β-catenin regulation [15]. The maintenance of a balance between the systemic calcium homeostasis and the biomechanical properties of bone tissue is another example of the crucial function performed by Wnt signaling. Loss-of-function mutations in Wnt/β-catenin signaling components such as LRP5 and SOST have led to severely decreased or increased bone mass and osteoporosis [16]. Interestingly, an increase in bone mass is associated with decreased activity of Wnt inhibitors such as Dickkopf-1 or Sclerostin. This data showed that Wnt signaling has a substantial role in bone growth [17].

In Gurley et al.'s study to understand the mechanisms of anterior-posterior differentiation, β-catenin was suggested as a molecular switch during regeneration and homeostasis in planarians [18]. Regarding metabolic homeostasis, attention to the association of TCF7L2 polymorphisms with type 2 diabetes has emphasized the function of the canonical Wnt pathway in glucose homeostasis [19]. Furthermore, new research indicates that skin homeostasis requires an appropriate level of Wnt/β-catenin signaling. The high grade of Wnt signaling directs keratinocytes to form hair, whereas low or intermediate levels of Wnt signaling determine interfollicular epidermis and the sebocyte lineages. Disturbing the balance of Wnt/β-catenin signaling can lead to the excessive formation of particular skin cell lineages and cancer [20].

The equilibrium between biomaterial synthesis and degradation is essential to life. One of the most catabolic in eukaryotic cells is autophagy which plays a critical function in preserving stability by eliminating destroyed or aged organelles and toxic protein aggregates. So, autophagy is essential in cellular homeostasis by maintaining bioenergetics and establishing a critical response to microenvironmental stress. Current literature illustrates that autophagy and Wnt/β-catenin signaling are related and have cross-talk at different stages of cellular development and differentiation to maintain cellular homeostasis. During nutritional deficiencies, β-catenin and Disheveled become targets for degradation by microtubule-associated protein light chain 3 (LC3) as an autophagic degradation protein. The Wnt signaling pathway by β-catenin, a corepressor of autophagy proteins of p62 and GSK3β, could regulate Wnt and control autophagy adversely [21].

3. Key proteins of the pathway

The original Wnt signaling pathway in simple metazoans, including *Porifera* (sponges), *Placozoa*, and *Ctenophora* (comb jellies), is made up of a basal number of transducers, including Wnt molecules, receptors, and cytoplasmic [22]. Despite variations in details of Wnt ligands, receptors, and cytoplasmic proteins involved in signaling pathways, they shared the main components, such as extracellular ligands and membrane receptors. Since The Wnt/−βcatenin pathway, also known as the canonical pathway, is one of the main pathways in the renewal, proliferation, and differentiation of stem cells and adult tissue homeostasis, in the following, we will introduce the essential proteins of this pathway.

3.1 The Wnt family

Wnts (Wingless-type MMTV integration site family) are a group of genes that encode Wnt-secreted glycoproteins. Beyond the metazoans, no genes encoding for Wnt ligands have been discovered. The number of Wnt gene subfamilies has increased from the first metazoans to cnidarians, which have a complete set of Wnt genes. However, the loss and gain of Wnt gene subfamilies have been traced in metazoan evolution [22].

3.1.1 Structure

The Wnt family, as the initiators of the Wnt signaling pathway, is comprised of 19 cysteine-rich proteins in mammals [21, 23], which share similarities in size (350–400 amino acids in length). Each includes an amino-terminal signal sequence with 22–24 cysteine residues, highly preserved during evolution, which are critical in Wnts for their function [24, 25]. Despite about 35% sequence identity among the majority of Wnt proteins, members of a subgroup (such as WNT3 and WNT3a) show more identity in sequence (58–83%) and have overlapping expression sites [26]. Among human Wnt proteins, only crystal structure of human Wnt3 (PDB ID:6AHY) (**Figure 2A**) has been characterized. The Wnt conservation analysis discovered that binding site regions of the Wnt to Frizzled receptor, named thumb and finger index, were highly conserved during evolution (**Figure 2B**). The result represents the critical role of the interacting Wnt protein with Frizzled in regulating the Wnt signaling pathway. In addition, comparing the

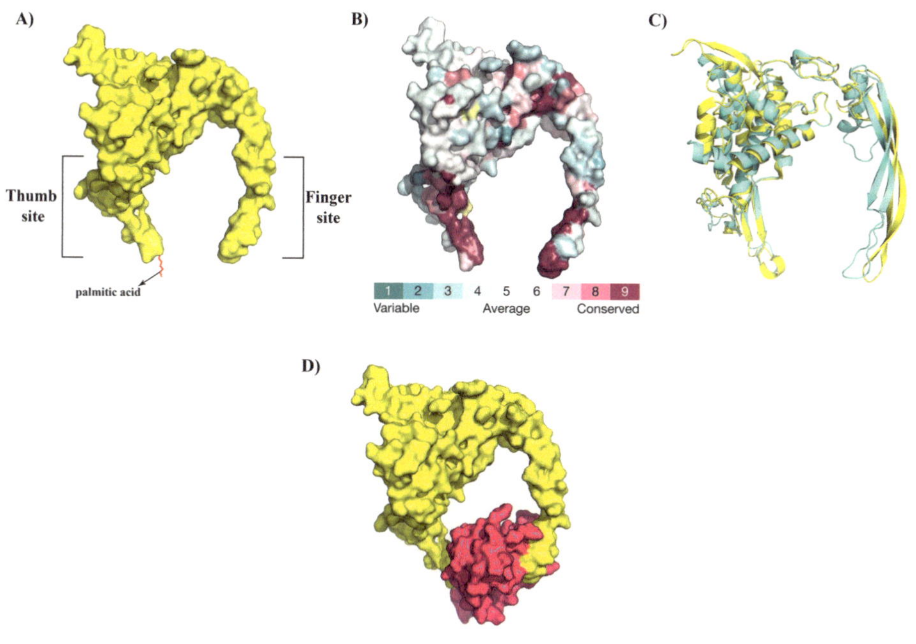

Figure 2.
The tertiary structure of human Wnt3. (A) Surface representation, (B) ConSurf webserver conservation analysis of WNT3. Data showed that binding site regions of WNT are conserved during evolution, (C) structure alignment of human Wnt3 (colored yellow) to Wnt8 Xenopus (colored cyan) shows the structure similarity of Wnts during evolution, and (D) complex of human Wnt3 (colored yellow) with the CRD domain (colored magenta) indicates two binding regions of Wnt to frizzled.

structure of human Wnt3 with that of Wnt8 *Xenopus* revealed the tertiary structure of Wnts conserved throughout evolution (**Figure 2C**).

Because of the existence of O-lipidation at a conserved serine, Wnt proteins are highly hydrophobic. The first Wnt protein structure (complex of *Xenopus* Wnt8 with mouse Frizzled 8) revealed that Wnts attach to Frizzled at two different sites on opposite faces of the cysteine-rich domain (**Figure 2D**) (CRD) [27]. This conserved extracellular domain in all Frizzled proteins contains 10 cysteine residues that bind to multiple Wnts with high affinity [28].

The Wnt family can be categorized according to the features and functions of Wnt1 and Wnt5a. The canonical signaling pathway comprises the Wnt1 group, including Wnt1, Wnt2, Wnt2b, Wnt3, Wnt3a, Wnt7a, Wnt8, Wnt8b, and Wnt10a components. While Wnt4, Wnt5a, and Wnt11 are all members of the Wnt5a family, they have the potential to stimulate a noncanonical signaling pathway [29].

3.1.2 Role and function

In all Wnt signaling pathways, Wnt molecules interact with their Frizzled or co-receptors to initiate a signaling cascade. The canonical Wnt pathway begins by binding Wnt molecules to the Frizzled receptor and LRP5/6. Also, Wnt ligands in the category of Wnt5a could stimulate the PCP and Wnt/Ca2+ pathways, which are characterized as noncanonical pathways. Signaling is transferred via Frizzled binding and DVL activation without LRP as a co-receptor [12].

3.1.3 Evolution of Wnts

Wnt ligands are an ancient and diverse protein family that induces target cells by binding to Frizzled receptors [4]. Wnt genes are identified from sponges to humans that become 13 subfamilies (Wnt1 to Wnt11, Wnt16, and WntA) by duplication and diversification before splitting the bilaterian and cnidarian. Phylogenetic and genomic investigations on a large scale have discovered that some Wnt genes are lost and retained during evolution in animals. This data means that each lineage of animals constructs its reservoir of Wnt genes [30]. Also, phylogenetic analysis of the orthologues of Wnts during evolution demonstrates high sequence similarity; for example, the human Wnt1 has 98% identity to the mouse Wnt1, which means evolutionary relationships between these proteins during development [31]. According to the structural analysis of Wnt proteins during evolution, the thumb site of Wnt proteins is closely related to saposin-like proteins, a primitive category of multifunctional lipid-interacting and helical carrier folds. In contrast, the finger site of Wnt proteins resembles a cytokine-like domain. During evolution, these two domains of Wnt came together to make more specific and robust interactions with Frizzled receptors to control animal development and homeostasis [32].

3.2 The SFRP family

Numerous antagonists regulate Wnt signaling, such as Wnt inhibitory factor 1 (WIF1), Cerberus, Sclerostin, members of the Dickkopf, and secreted Frizzled-related protein (SFRP) families. Sclerostin and Dickkopf proteins bind to LRP5 or LRP6 and block canonical signaling, whereas WIF1, Cerberus, and SFRPs can interact directly with Wnt proteins and inhibit their activation [33, 34].

The most prominent family of Wnt inhibitors is the SFRPs. This family is identified in all vertebrates and comprises five secreted glycoproteins: SFRP1, SFRP2, SFRP3 (Frzb), SFRP4, and SFRP5 [35]. Sequence and phylogenetic analysis illustrate that the SFRPs family consists of two subgroups: SFRP1, SFRP2, and SFRP5, which are strongly linked, and SFRP3 and SFRP4, which cluster together. This classification reveals a distinct organization of SFRPs' genomes. Surprisingly, the third subgroup of SFRPs, which appears to be absent in mammals, has been discovered in *Xenopus*, zebrafish, and chicks named "Sizzled," "Crescent," and "TLC," respectively. This group shows sequence similarity with the SFRP1-SFRP2-SFRP5 subgroup with an additional inter-domain disulfide bond in the Sizzled and Crescent subgroups [17].

3.2.1 Structure

The SFRP family consists of a CRD with 10 cysteine residues at the N-terminus that shares 30–50% sequence homology with the CRD of the Frizzled protein and a netrin-related motif (NTR) at the C-terminus that folds into two independent domains (**Figure 3A**) [35]. It is presumed that the CRD domain is composed mainly of the alpha helix [37] and interacts with the Wnt in a similar interface mode to the Frizzled CRD domain [38–40]. The NTR domain in the SFRP family has approximately 120 amino acids with six cysteine residues. Compared to the CRD domain, the cysteines in this domain are not conserved during evolution [41]. The NTR domain displays homology with netrins, complement C3, C4, C5, and procollagen C-endopeptidase enhancers [42]. The tertiary structure of this domain comprises five beta barrels [43–45]. The role of the NTR domain in SFRPs is not yet fully understood [46]. However, Bhat et al. revealed that both parts of SFRPs are critical for their particular inhibitory activity [47]. Some studies have illustrated that the NTR domain regulates the Wnt signaling

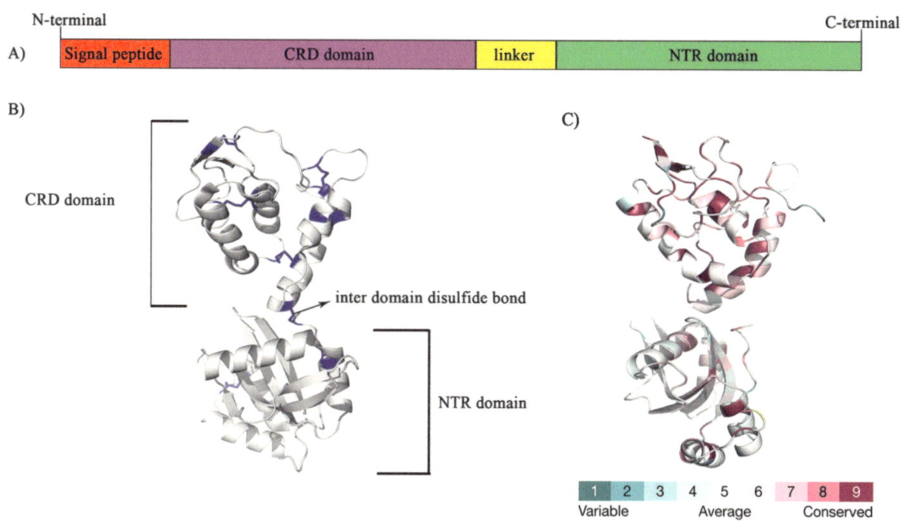

Figure 3.
Schematic view of sequence and structure analysis of sizzled protein. (A) Sequence analysis of sizzled protein with its different regions, (B) tertiary structure of the sizzled protein (PDBID:5XGP) [36]. Disulfide bonds in a structure characterized by the blue stick. Inter-domain disulfide bond has just existed in sizzled protein and not in SFRPs, and (C) conservation analysis of sizzled during evolution by ConSurf webserver. Data show the CRD domain is more conserved than the NTR domain during evolution.

pathway by binding to heparin [48, 49] via its positive charge of lysine and arginine amino acids located at the bottom of the domain, contrary to the NTR-CRD interface [36]. Also, Bhat et al. revealed that any alteration of the lysine residues to alanine in the NTR domain of SFRP1 caused a reduction of antagonist activity in the Wnt signaling pathway [47]. The only crystal structure of full-length SFRPs belongs to the Sizzled protein of *Xenopus laevis* (**Figure 3B**) as a homolog protein of SFRPs [36]. Conservation analysis showed that the CRD domain of Sizzled was more conserved than the NTR domain due to its critical role in interaction with Wnts (**Figure 3C**). Moreover, SFRPs represent posttranslational modifications in their structure. This alteration seems to confer additional differences which may cause further diversitification of various SFRP family functions, like N-glycosylation of SFRP1 and sulfated at two tyrosine residues in SFRP5, which inhibit its binding to heparin and reduce the stability of SFRP5 [17].

3.2.2 Role and function

The SFRP gene family is regarded as a tumor suppressor. However, new studies indicate that SFRPs could play multiple roles in regulating the Wnt signaling pathway and cause various effects in various kinds of cancer due to changes in their expression [35]. Although SFRP has been identified as a critical Wnt signaling regulator, the precise mechanism by which Wnt signaling is controlled remains unknown [17]. Moreover, SFRPs have been linked to disorders such as skeletal diseases, ocular degradation, and hypophosphatemic illnesses, implying that their activity is required for tissue homeostasis [17]. Phosphorus is essential in a variety of biological processes. Phosphorus homeostasis is linked to various cellular mechanisms involved in energy homeostasis, signal transduction, synthesis of nucleic acids, cellular membrane function, bone wellness, and integrity. Therefore, maintaining phosphorus balance and homeostasis is essential to the organism's health. The SFRP4, characterized as phosphatonin molecules, a new class of hormones that regulate renal phosphorus wasting, plays a crucial role in phosphorus homeostasis by antagonizing the Wnt pathway in the kidney. SFRP4 overexpression has been found in cancers associated with wasting phosphate and osteomalacia in the kidney. In both events, the abundance of Na + -Pi symporters decreases in the proximal tubule of the brush border membrane. This reduction caused a decrease in the expression of the Na + -Pi IIa symporter at the surface of the proximal tubules and opossum cells in the renal. Also, subsequent studies illustrate that this protein participates in other pathophysiological circumstances related to the wasting of phosphate, which could contribute to physiological regulation [50, 51]. Moreover, SFRP1 could also affect this pathway by interaction with RankL. Likewise, SFRP3 is related to the progress of osteolysis or heterotopic ossification [17].

Alternately, SFRPs have an impact on promoting and generating photoreceptors during the development of embryos. Elevated SFRPs expression, especially SFRP1, caused retinal damage in patients with retinitis pigmentosa, a congenital disorder characterized by the promotion of photoreceptor loss [17].

3.2.3 Evolution of SFRPs

SFRP homologs have not been discovered in the *Drosophila* genome. At the same time, this family exists in other invertebrates, such as the purple sea urchin, the nematode *Caenorhabditis elegans*, and even the sponge *Lubomirskia baikalensis*, demonstrating the SFRPs' historical basis (4). This organism only has one SFRP protein, whereas most invertebrates (except *lamprey*) have two SFRP proteins and all vertebrates have five SFRP proteins. This data demonstrated how SFRP proteins duplicated and diverged during the evolution of vertebrates to more precisely and accurately control signaling pathways [52]. Moreover, sequence analysis of SFRP protein in humans found that disulfide bonds in the CRD domain are completely conserved while not in the NTR domain. This result could be interpreted since the origin of the CRD domain of human SFRPs came about by the duplication of CRD domains of Frizzled receptors during evolution. Still, the NTR domain has been independently created [53].

3.3 The DKK family

The DKK (Dickkopf) protein family is a group of four Wnt modulators in vertebrates, with sizes ranging from 255 to 350 amino acids. The distinct feature of these proteins is two conserved CRDs that differ from the CRDs found in SFRPs and Frizzleds [54].

3.3.1 Structure

The Dickkopf protein in vertebrates is composed of two conserved CRD regions(CRD1 and CRD2), which are separated by a spacer region of different lengths in DKK members (**Figure 4A**). CRD2 competes with Wnt ligands for LRP5/LRP6 interaction, whereas CRD1 modulates CRD2 [55]. In their CRD1 and CRD2 domains, DKK1 and DKK2 share 50 and 70% of their identity, respectively. The CRD2 domains of DKK1 and DKK2 have a high affinity for direct interaction with LRP 5/6. The DKK2 (CRD2) NMR structure revealed a relatively flat molecule with six-sheet regions composed of five disulfide bonds, stabilizing the CRD2 structure [56]. Furthermore, DKK3 is unique within the Dickkopf family in that it contains an N-terminal Soggy

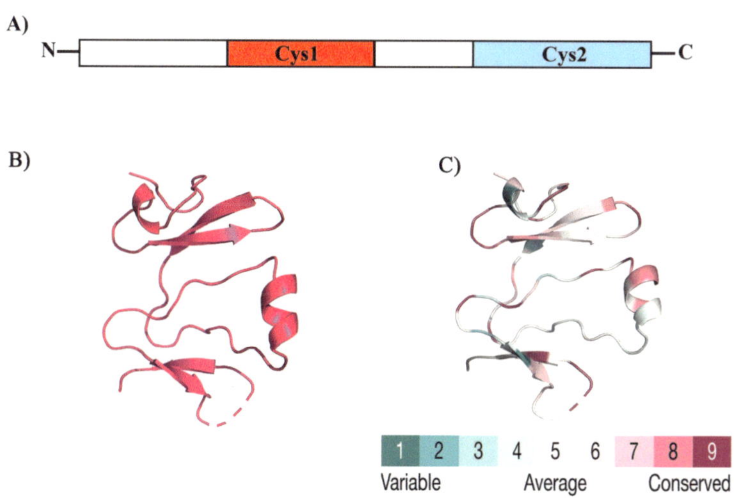

Figure 4.
Schematic view of the human DKK1 domain and the 3D structure of the human DKK1 CRD2 domain. (A) DKK domains, (B) tertiary structure of DKK1's CRD2 domain (PDB ID: 3SOQ), which interacts with the LRP receptor, and (C) conservation analysis of the CRD2 domain throughout evolution.

domain as well as two CRD domains [57]. The tertiary structure of the CRD2 domain of human DKK1 is represented in **Figure 4B**. Conservation analysis of this domain illustrated that this domain is approximately highly conserved during evolution (**Figure 4C**).

3.3.2 Role and function

Generally, Dickkopf proteins are critical for embryogenesis and osteogenesis. So, any abnormalities in DKK protein have been linked to a range of cancers, bone diseases, and neurodegenerative disorders [56]. DKK1 binds to LRP5/LRP6 and its co-receptor Kremen to block the Wnt pathway, which is needed for the development of the head and limb formation in vertebrate embryos. Dickkopf proteins were involved in one of the negative-feedback loops that terminated or repressed activation of Wnt at the cell surface receptor. Indeed, DKK1 is a promising target gene of the Wnt/β-catenin pathway that inhibits the initiation of signaling by competing with Wnts to interact with LRP5/6 [58, 59].

Wu et al. discovered that DKK2 and Wnt collaborate synergistically to stimulate the Wnt/β-catenin pathway in embryos of Xenopus. It seems that the function of DKK2, disparate from other members of the DKK family, appears to be a stimulated co-factor for Wnt signaling [55, 60]. DKK3, on the other hand, unlike the other DKKs, is not a Wnt signaling antagonist [57].

3.3.3 Evolution of DKKs

Although the homologs of the receptor proteins Frizzled and LRP5/LRP6 have been identified in vertebrates and insects, no Dickkopf protein was discovered in the genomes of insects and nematodes. By finding a DKK3-related protein in *Hydra*, DKK3 was proposed as the ancestral Dickkopf type. Structural and phylogenetic analysis indicated that the vertebrate DKK1/2/4 subfamily was created by subsequent gene duplication. Furthermore, functional analyses of HyDkk1/2/4 and canonical Wnts in *Xenopus* and *Hydra* proposed that the Wnt-Dickkopf antagonism existed in the last common origin of cnidarians and bilaterians, while it was lost in Caenorhabditis and insects [61].

3.4 The LRP family

Low-density lipoprotein receptors (LDLRs) are a protein family that is found in a variety of tissues and have a variety of cellular functions. LRP5 and LRP6 are Wnt ligand co-receptors in canonical Wnt signaling that significantly influence various physiological and pathological procedures in adult tissues. The dysregulation of LRP5 and LRP6 has been associated with various diseases ranging from bone, cardiac, neurodegenerative, diabetes, and hypercholesterolemia to cancer. Though impaired LRP6 signaling is associated with human diseases, studies suggest a predominant role of LRP6 during development, where loss-of-function of LRP6 in mice was embryonically lethal [62–65].

3.4.1 Structure

The LRP5/6 are single-pass transmembrane receptors with an extracellular domain that comprises four tandem YWTD-type b-propeller (BP)-epidermal growth factor (EGF)-like domains, which are named P1E1-P2E2, and P3E3–P4E4. These regions are followed by three LDLR type (A) repeats (**Figure 5**). The extracellular regions of LRP5/6 are responsible for interacting with Wnt molecules and antagonists, including Dickkopf-related protein 1 (DKK1) and Sclerostin [66]. The domains of P1E1-P2E2 primarily interact with Wnt1, Wnt2, Wnt2b, Wnt6, Wnt8a, Wnt9a, Wnt9b, and Wnt10b, while Wnt3 and Wnt3a bind to P3E3-P4E4 domains. The cytoplasmic region of LRP5/6 comprises 200 amino acids and five signature PPPSPxS motifs labeled A through E. The cytoplasmic domains of LRP5/6 have a binding site for Axin, which is required for LRP6 signaling. Though these proteins have been highly conserved during evolution, the amino acid sequences of LRP5/6 proteins share 71% identity [67]. The crystal structure of LRP5 has not been assessed yet. Although the whole structure of LRP6 does not exist, the structures of functional domains of LRP6, which interact with Wnt proteins, exist in the PDB bank (**Figure 6A** and **C**). Also, conservation analysis demonstrated that functional domains of LRP6 were nearly conserved during evolution (**Figure 6B** and **D**).

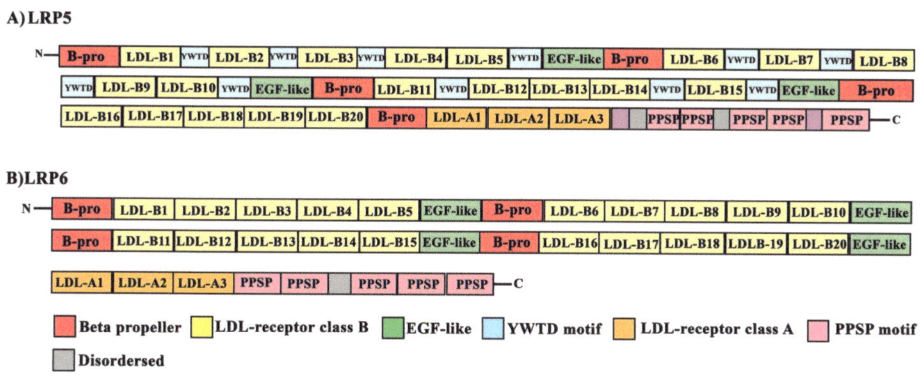

Figure 5.
Schematic view of LRP5/6 domains and regions in humans.

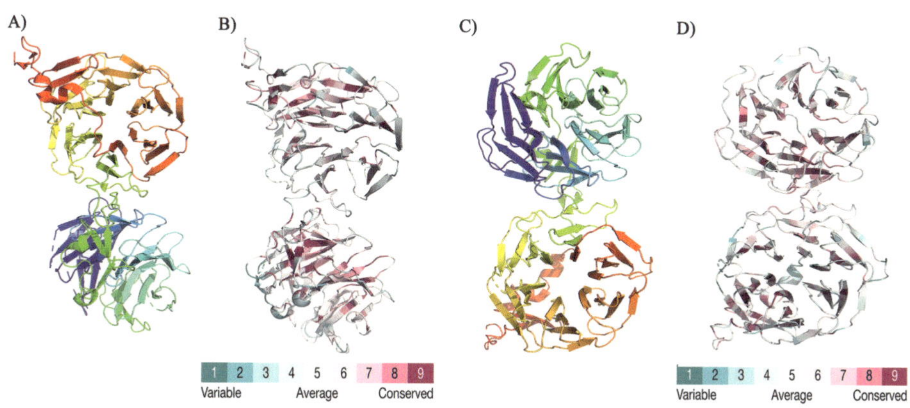

Figure 6.
The tertiary structure of LRP6. (A) the crystal of domains E1 and E2 of LRP6 (PDB ID: 3S94), (B) ConSurf web server conservation analysis of E1 and E2 regions of LRP6, (C) the crystal of domains E3 and E4 of LRP6 (PDB ID: 3S8Z), and (D) ConSurf web server conservation analysis of E3 and E4 domains of LRP6.

3.4.2 Role and function

LRP5 and LRP6 are co-receptors for canonical Wnt signaling, though LRP6 is more active than LRP5 in the Wnt canonical pathway. Binding the Wnt ligand to the cell surface receptors Frizzled (FZD) and LRP5/6 induces phosphorylation of the intracellular domain of LRP5/6, dissociation of the destruction complex, stabilization of cytosolic β-catenin, translocation of β-catenin into the nucleus, and eventually, activation of Wnt target gene transcription [68]. Wnt ligands are classified according to their preference for binding to LRP5 or LRP6. Wnt3, Wnt3A, Wnt6, Wnt8A, and Wnt8B only activate Wnt signaling through LRP6, whereas Wnt2 primarily promotes Wnt signaling through LRP5. Wnt1, Wnt7B, and Wnt9A interact similarly with LRP5 and LRP6 [66].

Several proteins, such as DKK1, Sclerostin, PEDF, and Kallistatin, inhibit Wnt binding by interacting with LRP6. However, besides the Wnt ligands, proteins such as the R-Spondin (Rspo) family, Norrin, parathyroid hormone (PTH), transglutaminase 2 (TG2), Cripto-1, Biglycan, and single-span membrane protein TMEM59 stimulate the Wnt pathway by interacting with LRP6. These findings demonstrate that LRP5/6 may also operate as the co-receptor of other ligands in initiating the Wnt/β-catenin signaling [66].

Although LRP6 expression varies between human tissue types, LRP6 is essential in regulating cell proliferation, differentiation, migration, and stem cell homeostasis. Regarding the unique function of LRP5/6 as the co-receptor for several extracellular cues, LRP5/6 has been an attractive target for designing a drug to abrogate/activate the Wnt signal pathway in recent years (**Table 1**) [69].

3.4.3 Evolution of LRP5/6

Several studies showed conservation evolutionarily in the Lrp5/6 gene among all animals. Investigation of the LDLR family in *Drosophila* indicated that the arrow gene, which defines LDL-receptor-related protein as a required protein for signaling

Company/University	Category	Targeting mechanism	Disease
Surrozen INC, USA	Antibodies	LRP5/6	Melanoma cancer
Yale University, USA	Antibodies	Interaction between DKK2 and LRP5	
Agency for Science, Technology, and Research (A*STAR), Singapore	Small molecule	phosphorylated LRP6 (Ser1490)	Human pancreas HPAF-II cancer cells
Boehringer Ingelheim international GMBH, Germany	Polypeptides	LRP5/6	Cancer
Nantbio Inc., USA	Small molecule	Phosphorylation of LRP6	—
Surrozen INC, USA	Antibody	LRP6	Osteoporosis
Surrozen INC, USA	Antibody	LRP6	Retinopathy

Table 1.
List of released patents for targeting LRP6 in various diseases.

in the Wingless pathway, is structurally similar to LRP5 and LRP6 in vertebrates. Despite having 45% homology with the LRP5 and LRP6 proteins, arrow exhibits phenotypic similarities with Wg/Wnt mutants. Furthermore, there is a 98% amino acid sequence similarity between human and mouse LRP6. Furthermore, at the protein level, LRP6 shares more than 93% identity with humans and chickens [63].

3.5 The frizzled family

The Frizzled (FZD) family is characterized as G-protein-coupled receptors (GPCRs) that contribute significantly to embryogenesis. According to robust evidence, this family controls tissue homeostasis in various adult organs [70]. Any dysregulation of this family causes multiple types of disease in adults or during embryonic development. These problems include cancer, cardiac hypertrophy, familial exudative vitreoretinopathy, brain synapses, tissue and cell polarity, proliferation control, and schizophrenia in human and animal models [70, 71]. This family could activate three main signaling pathways: the FZD/β-catenin pathway, the FZD/Ca2+ pathway, and the FZD/PCP (planar cell polarity) pathway. Many secreted ligands, including all forms of Wnt protein, the SFRP family, R-Spondin, and Norrin, could interact directly with Frizzled and initiate the downstream signaling pathway [70].

Based on sequence analysis, humans have 10 Frizzled genes, comprising four primary classes. According to sequence classification, FZD1, FZD2, and FZD are classified into one group, which has 75% of the amino acids are identical; the second group consists of FZD5 and FZD8 with 70% amino acid identity to each other; the third group consists of FZD4, FZD9, and FZD10 with 65% identical amino acids in sequences. The last group is FZD3 and FZD6, sharing 50% amino acid identity [71].

3.5.1 Structure

Frizzled proteins (FZD) have different lengths ranging from 500 to 700 amino acids. They are composed of an extracellular region at the N-terminus, comprising a cysteine-rich domain (CRD), a hydrophilic linker region of 40–100 amino acids, and a transmembrane area is compromised of the seven-pass transmembrane domain receptors in alpha helix form. The intracellular domain at the C-terminus region is highly variable among different Frizzled family members (**Figure 7A**) [71]. The CRD domain of Frizzled protein is responsible for its interaction with Wnt protein, consisting of 120–125 residues and a fully conserved disulfide bond. The only crystal structure of Frizzled 2 (PDB ID: 6C0B), 4 (PDB ID: 5CM4), 5 (PDB ID: 5039), 7 (PDB ID: 5 T44), and 8 (PDB ID: 5UN5) of humans has been discovered yet belongs to the CRD domain. The CRD region crystal structure analysis of Frizzled proteins revealed that the dominant structure of this region is the alpha helix, which is highly conserved during evolution (**Figure 7B**) [37]. Conservation analysis of the human Frizzled CRD region showed that this domain is highly conserved during evolution (**Figure 7C**). Sequence similarity search reveals that conservation of transmembrane domain of Frizzled compared to other GPCR is low and limited to the hydrophobic residues. According to sequence similarity search, this family might be the evolution from the Taste2 subfamily of taste receptors. After the transmembrane domain, the highly conserved KTXXXW motif is essential for the Wnt/β-catenin pathway activation. Apart from this motif, other regions in the carboxy-terminus domain are not utterly conserved among Frizzleds [71].

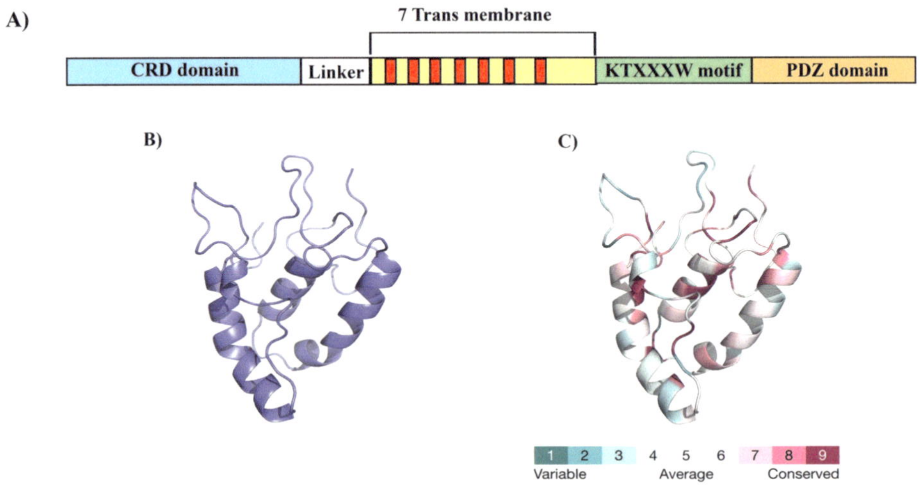

Figure 7.
Schematic view of the frizzled sequence and tertiary structure analysis of the CRD domain. (A) Frizzled protein domains, (B) tertiary structure of frizzled 4's CRD domain (PDB ID: 5CM4), revealing that most of the CRD domain consists of the alpha structure, and (C) conservation analysis of the CRD domain by the ConSurf web server reveals that this domain was highly conserved during evolution.

3.5.2 Role and function

Frizzled proteins play an essential role in the Wnt signaling pathway. This protein family controls various signaling pathways, including PCP, canonical Wnt, and Wnt-calcium signaling, which could be activated by Wnt ligands interacting with Frizzled receptors. Targeted mutation analysis in the Frizzled family illustrates this family's role in a vast range of development and homeostatic processes. This process includes morphogenetic movements responsible for the palate, ventricular septum, ocular furrow, and neural tube closure; the existence of thalamic neurons; osteogenesis; angiogenesis of the central nervous system (CNS); generation and conservation of the blood-brain barrier; and a huge range of procedures that are responsible for cellular, subcellular, and multicellular orientation constructions that are related to the body's orientations [72]. Also, upregulated FZD receptor expression in several cancer malignancies affects patient outcomes (survival and recurrence), followed by activation of the Wnt signaling pathway [73].

3.5.3 Evolution of frizzled

The Frizzled genes were discovered in Drosophila while looking for mutations that could affect the polarity of adult fly epidermal cells. Later, Frizzled proteins were found in various metazoans but not in protozoans. The early metazoans, such as the sponge *Suberites domuncula* and *Hydra vulgaris* have Frizzled genes. *Caenorhabditis elegans* as a roundworm and *Drosophila* have three and four Frizzled genes, respectively. At least 10 Frizzled genes have been identified in vertebrates [71]. Sequence analysis of Frizzled genes in various species shows that around 20–40% of their sequence is identical. Furthermore, genomic organization analysis of Frizzled genes in humans and invertebrates shows that Frizzled genes do not appear to be evolutionarily conserved across a wide range of species, such as the lack of introns region in human FZD1, FZD2, and FZD7 to FZD10, whereas other FZD genes, such as human

FZD5, have one intron like Drosophila Frizzled 2 (Dfz2). Even though the lack of introns in Frizzled genes seems to happen in *Drosophila*, it also occurs in humans. This data shows that Frizzled genes are derived from a common ancestor [71].

4. Conclusion

Though the three Wnt signaling pathways are initiated by the identical initial events of interacting Wnt molecules with Frizzled receptors, various proteins are involved in providing the specificity of function downstream that defines each pathway. The canonical pathway has been broadly studied as a significant factor in animal life, from protists to vertebrates, to achieve tissue homeostasis and development. In contrast, limited knowledge is available on the evolutionary distribution and the molecular evolutionary preservation of the Wnt/PCP and the Wnt/calcium pathway components during evolution.

Author details

Elham Rismani[1], Nasrin Haghighi-Najafabadi[1,2], Babak Elyasi Far[3], Behzad Shahbazi[1] and Ladan Mafakher[4*]

1 Molecular Medicine Department, Biotechnology Research Center, Pasteur Institute of Iran, Tehran, Iran

2 Virology Department, Pasteur Institute of Iran, Tehran, Iran

3 Department of Physiology and Pharmacology, School of Medicine, Dezful University of Medical Sciences, Dezful, Iran

4 Thalassemia and Hemoglobinopathy Research Center, Health Research Institute, Ahvaz Jundishapur University of Medical Sciences, Ahvaz, Iran

*Address all correspondence to: ladan.mafakher@gmail.com

References

[1] Srivastava M, et al. The Amphimedon queenslandica genome and the evolution of animal complexity. Nature. 2010;**466**(7307):720-726

[2] Nusse R, Varmus HE. Many tumors induced by the mouse mammary tumor virus contain a provirus integrated in the same region of the host genome. Cell. 1982;**31**(1):99-109

[3] Rao TP, Kühl M. An updated overview on Wnt signaling pathways: A prelude for more. Circulation Research. 2010;**106**(12):1798-1806

[4] Janssen R et al. The evolution and expression of panarthropod frizzled genes. Frontiers in Ecology and Evolution. 2015;**3**:96

[5] Nusse R, Clevers H. Wnt/β-catenin signaling, disease, and emerging therapeutic modalities. Cell. 2017;**169**(6):985-999

[6] Perugorria MJ et al. Wnt–β-catenin signalling in liver development, health and disease. Nature Reviews Gastroenterology & Hepatology. 2019;**16**(2):121-136

[7] Kitagawa M et al. An F-box protein, FWD1, mediates ubiquitin-dependent proteolysis of β-catenin. The EMBO Journal. 1999;**18**(9):2401-2410

[8] Aberle H et al. β-Catenin is a target for the ubiquitin–proteasome pathway. The EMBO Journal. 1997;**16**(13):3797-3804

[9] Liu J et al. Wnt/β-catenin signalling: Function, biological mechanisms, and therapeutic opportunities. Signal Transduction and Targeted Therapy. 2022;**7**(1):3

[10] Seifert JRK, Mlodzik M. Frizzled/PCP signalling: A conserved mechanism regulating cell polarity and directed motility. Nature Reviews Genetics. 2007;**8**(2):126-138

[11] Gordon MD, Nusse R. Wnt signaling: Multiple pathways, multiple receptors, and multiple transcription factors. The Journal of Biological Chemistry. 2006;**281**(32):22429-22433

[12] Komiya Y, Habas R. Wnt signal transduction pathways. Organogenesis. 2008;**4**(2):68-75

[13] Sugimura R, Li L. Noncanonical Wnt signaling in vertebrate development, stem cells, and diseases. Birth Defects Research Part C: Embryo Today: Reviews. 2010;**90**(4):243-256

[14] Petersen CP, Reddien PW. Wnt signaling and the polarity of the primary body axis. Cell. 11 Dec 2009;**139**(6):1056-1068

[15] Clevers H, Loh Kyle M, Nusse R. An integral program for tissue renewal and regeneration: Wnt signaling and stem cell control. Science. 2014;**346**(6205):1248012

[16] Steinhart Z, Angers S. Wnt signaling in development and tissue homeostasis. Development. 2018;**145**(11):dev146589

[17] Bovolenta P et al. Beyond Wnt inhibition: New functions of secreted frizzled-related proteins in development and disease. Journal of Cell Science. 2008;**121**(6):737-746

[18] Gurley KA, Rink JC, Sánchez Alvarado A. Beta-catenin defines head versus tail identity during planarian regeneration and homeostasis. Science. 2008;**319**(5861):323-327

[19] Jin T. Current understanding on role of the Wnt signaling pathway effector TCF7L2 in glucose homeostasis. Endocrine Reviews. 2016;**37**(3):254-277

[20] Lim X, Nusse R. Wnt signaling in skin development, homeostasis, and disease. Cold Spring Harbor Perspectives in Biology. 2013;**5**(2):1-25

[21] Lorzadeh S et al. Autophagy and the Wnt signaling pathway: A focus on Wnt/β-catenin signaling. Biochimica et Biophysica Acta (BBA)-molecular. Cell Research. 2021;**1868**(3):118926

[22] Holstein TW. The evolution of the Wnt pathway. Cold Spring Harbor Perspectives in Biology. 2012;**4**(7):1-17

[23] Bodine PV. Wnt signaling control of bone cell apoptosis. Cell Research. 2008;**18**(2):248-253

[24] Du SJ et al. Identification of distinct classes and functional domains of Wnts through expression of wild-type and chimeric proteins in Xenopus embryos. Molecular and Cellular Biology. 1995;**15**(5):2625-2634

[25] van Ooyen A, Kwee V, Nusse R. The nucleotide sequence of the human int-1 mammary oncogene; evolutionary conservation of coding and non-coding sequences. The EMBO Journal. 1985;**4**(11):2905-2909

[26] Miller JR. The Wnts. Genome Biology. 2002;**3**(1):REVIEWS3001

[27] Janda CY et al. Structural basis of Wnt recognition by frizzled. Science. 2012;**337**(6090):59-64

[28] Mikels AJ, Nusse R. Wnts as ligands: Processing, secretion and reception. Oncogene. 2006;**25**(57):7461-7468

[29] He S et al. Wnt3a: Functions and implications in cancer. Chinese Journal of Cancer. 2015;**34**(3):1-9

[30] Somorjai IM et al. Wnt evolution and function shuffling in liberal and conservative chordate genomes. Genome Biology. 2018;**19**(1):1-17

[31] Miller JR. The wnts. Genome Biology. 2001;**3**(1):1-15

[32] Bazan JF, Janda CY, Garcia KC. Structural architecture and functional evolution of Wnts. Developmental Cell. 2012;**23**(2):227-232

[33] Kawano Y, Kypta R. Secreted antagonists of the Wnt signalling pathway. Journal of Cell Science. 2003;**116**(13):2627-2634

[34] Li X et al. Sclerostin binds to LRP5/6 and antagonizes canonical Wnt signaling. Journal of Biological Chemistry. 2005;**280**(20):19883-19887

[35] Liang C-J et al. SFRPs are biphasic modulators of Wnt-signaling-elicited cancer stem cell properties beyond extracellular control. Cell Reports. 2019;**28**(6):1511-1525. e5

[36] Bu Q et al. The crystal structure of full-length sizzled from Xenopus laevis yields insights into Wnt-antagonistic function of secreted frizzled-related proteins. Journal of Biological Chemistry. 2017;**292**(39):16055-16069

[37] Dann CE et al. Insights into Wnt binding and signalling from the structures of two frizzled cysteine-rich domains. Nature. 2001;**412**(6842):86-90

[38] Leyns L et al. Frzb-1 is a secreted antagonist of Wnt signaling expressed in the Spemann organizer. Cell. 1997;**88**(6):747-756

[39] Lin K et al. The cysteine-rich frizzled domain of Frzb-1 is required and sufficient for modulation of Wnt signaling. Proceedings of the National Academy of Sciences. 1997;**94**(21):11196-11200

[40] Wang S et al. Frzb, a secreted protein expressed in the Spemann organizer, binds and inhibits Wnt-8. Cell. 1997;**88**(6):757-766

[41] Chong JM et al. Disulfide bond assignments of secreted frizzled-related protein-1 provide insights about frizzled homology and netrin modules. Journal of Biological Chemistry. 2002;**277**(7):5134-5144

[42] Bányai L, Patthy L. The NTR module: Domains of netrins, secreted frizzled related proteins, and type I procollagen C-proteinase enhancer protein are homologous with tissue inhibitors of metalloproteases. Protein Science. 1999;**8**(8):1636-1642

[43] Fernandez-Catalan C et al. Crystal structure of the complex formed by the membrane type 1-matrix metalloproteinase with the tissue inhibitor of metalloproteinases-2, the soluble progelatinase a receptor. The EMBO Journal. 1998;**17**(17):5238-5248

[44] Muskett FW et al. High resolution structure of the N-terminal domain of tissue inhibitor of metalloproteinases-2 and characterization of its interaction site with matrix metalloproteinase-3. Journal of Biological Chemistry. 1998;**273**(34):21736-21743

[45] Tuuttila A et al. Three-dimensional structure of human tissue inhibitor of metalloproteinases-2 at 2.1 Å resolution. Journal of Molecular Biology. 1998;**284**(4):1133-1140

[46] Semënov MV, He X. Secreted Antagonists/Modulators of Wnt Signaling, in Madame Curie Bioscience Database [Internet]. Landes Bioscience. 2013

[47] Bhat RA et al. Structure–function analysis of secreted frizzled-related protein-1 for its Wnt antagonist function. Journal of Cellular Biochemistry. 2007;**102**(6):1519-1528

[48] Zhong X et al. Regulation of secreted frizzled-related protein-1 by heparin. Journal of Biological Chemistry. 2007;**282**(28):20523-20533

[49] Uren A et al. Secreted frizzled-related protein-1 binds directly to wingless and is a biphasic modulator of Wnt signaling. Journal of Biological Chemistry. 2000;**275**(6):4374-4382

[50] Berndt T, Kumar R. Phosphatonins and the regulation of phosphate homeostasis. Annual Review of Physiology. 2007;**69**:341-359

[51] Berndt T, Kumar R. Novel mechanisms in the regulation of phosphorus homeostasis. Physiology. 2009;**24**(1):17-25

[52] Yan J et al. The evolutionary analysis reveals domain fusion of proteins with frizzled-like CRD domain. Gene. 2014;**533**(1):229-239

[53] Bastin BR et al. Structure, phylogeny, and expression of the frizzled-related gene family in the lophotrochozoan annelid Platynereis dumerilii. EvoDevo. 2015;**6**(1):1-24

[54] Niehrs C. Function and biological roles of the Dickkopf family of Wnt modulators. Oncogene. 2006;**25**(57):7469-7481

[55] Li L et al. Second cysteine-rich domain of Dickkopf-2 activates canonical Wnt signaling pathway via LRP-6 independently of dishevelled. The Journal of Biological Chemistry. 2002;**277**(8):5977-5981

[56] Chen L et al. Structural insight into the mechanisms of Wnt signaling antagonism by Dkk. Journal of Biological Chemistry. 2008;**283**(34):23364-23370

[57] Baetta R, Banfi C. Dkk (Dickkopf) Proteins. Arteriosclerosis, Thrombosis, and Vascular Biology. 2019;**39**(7):1330-1342

[58] Semënov MV et al. Head inducer Dickkopf-1 is a ligand for Wnt coreceptor LRP6. Current Biology. 2001;**11**(12):951-961

[59] González-Sancho JM et al. The Wnt antagonist DICKKOPF-1 gene is a downstream target of β-catenin/TCF and is downregulated in human colon cancer. Oncogene. 2005;**24**(6):1098-1103

[60] Wu W et al. Mutual antagonism between dickkopf1 and dickkopf2 regulates Wnt/beta-catenin signalling. Current Biology. 2000;**10**(24):1611-1614

[61] Guder C et al. An ancient Wnt-Dickkopf antagonism in hydra. Development. 2006;**133**(5):901-911

[62] Singh R et al. LRP6 enhances glucose metabolism by promoting TCF7L2-dependent insulin receptor expression and IGF receptor stabilization in humans. Cell Metabolism. 2013;**17**(2):197-209

[63] Alrefaei AF, Abu-Elmagd M. LRP6 receptor plays essential functions in development and human diseases. Genes. 2022;**13**(1):120

[64] Nie X et al. LRP5 promotes cancer stem cell traits and chemoresistance in colorectal cancer. Journal of Cellular and Molecular Medicine. 2022;**26**(4):1095-1112

[65] Zhang L et al. LRP6 is involved in the proliferation, migration and invasion of trophoblast cells via miR-346. International Journal of Molecular Medicine. 2020;**46**(1):211-223

[66] Ren Q, Chen J, Liu Y. LRP5 and LRP6 in Wnt signaling: Similarity and divergence. Frontiers in Cell and Developmental Biology. 2021;**9**:1-11

[67] Kaur P et al. Membrane targeting of disheveled can bypass the need for arrow/lrp5. Scientific Reports. 2017;**7**(1):1-13

[68] Singh HD, Ma J-X, Takahashi Y. Distinct roles of LRP5 and LRP6 in Wnt signaling regulation in the retina. Biochemical and Biophysical Research Communications. 2021;**545**:8-13

[69] Goswami VG, Patel BD. Recent updates on Wnt signaling modulators: A patent review (2014-2020). Expert Opinion on Therapeutic Patents. 2021;**31**(11):1009-1043

[70] Schulte G, Bryja V. The frizzled family of unconventional G-protein-coupled receptors. Trends in Pharmacological Sciences. 2007;**28**(10):518-525

[71] Huang H-C, Klein PS. The frizzled family: Receptors for multiple signal transduction pathways. Genome Biology. 2004;**5**(7):1-7

[72] Wang Y et al. Frizzled receptors in development and disease. Current Topics in Developmental Biology. 2016;**117**:113-139

[73] Ueno K et al. Frizzled homolog proteins, microRNAs and Wnt signaling in cancer. International Journal of Cancer. 2013;**132**(8):1731-1740